The Biological Roots
of Human Nature

The Biological Roots of Human Nature

Forging Links between Evolution and Behavior

TIMOTHY H. GOLDSMITH

Andrew W. Mellon Professor of Biology
Yale University

New York   Oxford
OXFORD UNIVERSITY PRESS

Oxford University Press

Oxford New York Toronto
Delhi Bombay Calcutta Madras Karachi
Kuala Lumpur Singapore Hong Kong Tokyo
Nairobi Dar es Salaam Cape Town
Melbourne Auckland Madrid

and associated companies in
Berlin Ibadan

Goldsmith, Timothy H.
The biological roots of human nature: forging links between
evolution and behavior / Timothy H. Goldsmith.
p.   cm. Includes bibliographical references (p.   ) and index.
ISBN 0-19-506288-4
ISBN 0-19-509393-3 (pbk)
1. Sociobiology.   2. Social behavior in animals.   I. Title.
QH333.G65   1991
304.5—dc20   90-25281

3 5 7 9 8 6 4 2
Printed in the United States of America
on acid-free paper

For Margaret, who was,
And Margaret, who is and will be

Everything in the world has changed
except our way of thinking.
ALBERT EINSTEIN

Preface

Several years ago an economist colleague at Yale organized a series of informal faculty seminars to discuss the use of "evolutionary models and metaphors in the social sciences." The participants came from anthropology, biology, economics, law, linguistics, psychology, and sociology. We met weekly for lunch, and the discussions ranged widely. Although the larger intent of the organizers was to explore broadly the possible relevance to the social sciences of concepts that have arisen in the study of biological evolution, it soon became clear that an immediate stimulus had been the publication of E.O. Wilson's book *Sociobiology* and the ensuing publicity it had received. Consequently, much of the early discussion was focused on this topic.

As a biologist I was soon struck by the depth of skepticism that biology has anything interesting to say about the social behavior of humans. But it also seemed to me that social scientists, as a group, have a limited understanding of the rich fabric of evolutionary theory. On several occasions we were invited to read papers whose purported message was to expose the simplicity of "reductionist" biological thinking. The affect of the papers on me was quite different; invariably I found that the authors had simplistic views of biology in general and evolution in particular. How could such marvelously sweeping concepts be so thoroughly misunderstood? Increasingly, I found myself trying to articulate what had gone wrong by translating what I, as a biologist, took to be the meaning of evolutionary theory into words that my colleagues did not find ridiculous. Finally I tried to put it down on paper.

An earlier draft of the manuscript lay on my desk for several years. Each October I put several copies on reserve in the library as supplementary reading for my undergraduate course in neurobiology. This is a course about nerve cells and how knowledge of the behavior of neurons is helping us to understand the behavior of organisms. There is a point in the course where I encourage my students to contemplate several traditional but diverse approaches to the study of behavior—neurobiology, psychology, and behavioral evolution (with its roots in ethology)—traditions so disparate that they might have come from different planets. It is in this context that the manuscript has been made available to the class as optional reading, and each year I have been gratified that students have approached me to say that the scales have fallen from their eyes. This is the kind of reaction that sustains a teacher.

These two experiences reinforce my conviction that we do not teach science very effectively except narrowly to preprofessionals. For a wide variety of reasons, we do not do a good job of reaching students from other disciplines. This book is based on the premise that biology has a great deal to say that should be of interest to social scientists, historians, philosophers, and indeed humanists of all stripes. The science of biology has reached a point in its own development where it can begin to address some of the perennial questions about what it means to be human, themes that have been the province of literature and philosophy for centuries and supplemented by the social sciences in more recent times. I also write from the conviction that the contribution of biology to these issues can be conveyed in relatively simple language, without the camouflage of mathematics and with a minimum of jargon. I have also tried to do it without invoking more than the simplest reference to chemistry.

This, however, is not an apology for trivial ideas. The biological concepts in this book are deeply important for an understanding of what it means to be alive and to be a human being, but they will be new and challenging to many whose personal views about behavior, ethics, and societies have been shaped by very different traditions. For these readers the material may require considerable thought and reflection.

I am grateful to a number of colleagues who have read versions of the manuscript and who have been unstinting in their critical advice. Matters of both error and emphasis have been addressed as a result of this input. Richard Harrison was more than a sounding board for parts of Chapter 3, and his contribution to those sections might be better described as a closet collaboration. I also wish to thank John Bonner, Thomas Carew, Donald R. Griffin, Richard Nelson, David Polikansky, Elisabeth Vrba, Edward O. Wilson, William Zimmerman, and an anonymous reviewer, all of whom made trenchant suggestions, many of which have been followed. Mary Helen Goldsmith brought a critical eye and a fresh perspective to every page I put in front of her, and her encouragement has been sustaining. Sally Fisher and Beth Marks helped me to pursue references and offered their own opinions in the process. In spite of all this critical support, however, I reserve sole claim to any and all remaining errors of fact and lapses of judgment.

Woods Hole, Massachusetts   T.H.G.
August 1990

Contents

The Biological Roots
of Human Nature

1

The Dual Nature of Causation in Biology

Natural selection has built us, and it is natural selection we must understand if we are to comprehend our own identities. . . . Whole industries have grown up in the social sciences dedicated to the construction of a pre-Darwinian and pre-Mendelian view of the social and psychological world. . . . In short, Darwinian social theory gives us a glimpse of an underlying symmetry and logic in social relationships which, when more fully comprehended by ourselves, should revitalize our political understanding and provide the intellectual support for a science of medicine and psychology. In the process it should also give us a deeper understanding of the many roots of our suffering.   Robert Trivers[1]

[The concept of evolution] is a campaign of secularization in a scientific-materialistic society—a campaign to totally neutralize religious convictions, to destroy any concept of absolute moral values, to deny any racial difference, to mix all ethnic groups in cookbook proportions, and finally [to destroy] the differences between male and female.
    Unnamed author of "equal time" guidelines for biology textbooks in the State of California, quoted by Nelkin[2]

Most people do not consider themselves animals.
    Mel and Norma Gabler, as quoted by Schafersman[3]

These contrasting remarks, the first by an evolutionary biologist, the second two by citizens who have been active in efforts to influence the content of biology textbooks used in secondary schools, point to a sad and disturbing problem: More than a century after Charles Darwin called attention to one of the most exciting and unifying concepts in the whole of science, the teaching of evolution continues to be greeted with formidable resistance. In the entire sweep of scientific knowledge, only evolutionary biology is the object of political regulation in the public schools, justified by tireless and tiresome repetition of the words of William Jennings Bryan that, after all, evolution is "only a theory." Moreover, that after a half a century has passed a President of the United States can still utter the same phrase with conviction is testimony to the fact that the lay public's understanding of evolution remains in a primitive condition.

The issue has a deceptively crystalline form when cast as natural science versus the versions of creation found in Genesis. A literal reading of Genesis is not only out of joint with biological evidence, it is incompatible with physics, chemistry, geology, and astronomy—in short, with the whole of natural science. In many religious denominations this inconsistency is resolved by a symbolic or metaphorical reading of scriptures. Why then does the problem not recede? Why, for example, do politicians find a receptive audience when they castigate federal interference in education, referring to (among other things) the involvement of the National Science Foundation in the development of science curricula in the post–Sputnik years? Economic justification notwithstanding, one reason is that a proper study of evolution invites students to think in new ways about what it means to be a human being. For many people, this kind of intellectual curiosity is a threat to the moral and social order, and as suggested by the second of the quotations above, the teaching of evolutionary biology becomes not only a challenge to deeply held systems of belief, but a scapegoat for a variety of unsettling social problems as well. It is no accident that political passions are aroused, for most people desire more than a modicum of stability and predictability in their lives and are frequently made uncomfortable when their children adopt beliefs and take up life-styles that contradict what the parents feel is important.

The case can also be made, however, that biologists have largely failed to communicate with many of their colleagues in other disciplines, and that there is a rift of understanding that is just as deep and serious as the more public row with religious fundamentalists. As a test of this view I suggest that more than a few thoughtful readers will respond to the first of the quotations above with either bewilderment or outright antipathy.

This book illustrates some of the ways biology illuminates phenomena that are of interest to anyone who reflects on human affairs. Some of the biological themes that we will consider were brought to the fore about fifteen years ago under the banner of sociobiology. Unfortunately, sociobiology was politicized at the outset by those who saw in it an elaborate argument for justifying a competitive, capitalist status quo at the expense of a kinder, gentler, if socialist alternative. The legacy of Herbert Spencer's Social Darwinism dies hard. This brouhaha left the mistaken impression in some quarters that the biological issues raised by sociobiology could be accepted or dismissed on political or philosophical grounds. In an effort to defuse this ideological bomb, alternative terms for the subject have been proposed. One that captures the flavor for social scientists (albeit at the expense of biological breadth) is "evolutionary psychology." Whatever one chooses to call it, however, the infusion of biological knowledge is overdue if we are to construct a view of social systems and of behavior that is faithful to history as well as to the present. The message is particularly salient for the social sciences, because at the present time practitioners of different scientific disciplines sometimes seem to be speaking different languages.

Others have written syntheses, interpretations, essays, and textbooks about the evolutionary bases for behavior, and some of these are lucid and inspiring. What justification is there for yet another effort? One reason is emphasis. Most of the available accounts are built almost exclusively from evolutionary arguments about behavior that pay little heed to the nuts and bolts of how nervous systems work and come into being during development. I think this is a deficiency, because it leaves some of the arguments more abstract than they need to be, and I have tried to find a different balance.

I have assumed that the reader knows no more biology and chemistry than the nodding acquaintance acquired in secondary school, and I believe that with conscientious effort the concepts in this book can be understood by any intelligent person. A couple of geneticist friends (one female, one male) with whom I shared an earlier draft of the manuscript have independently told me, politely if slightly disdainfully, that I am simply describing common sense. That may be so for biologists, but experience tells me that a large number of other people will find otherwise. They will bring to their reading convictions about the world that will be difficult to harmonize with the perspective they find here. For this reason alone they will have to make a serious effort to understand.

The remainder of this chapter continues to set the stage, and the following two chapters are about evolution and evolutionary theory. Chapter 2 sets to rest several of the more egregious and silly ideas even educated people have about evolution; Chapter 3 tries to convey a bit of the complexity of contemporary evolutionary theory while at the same time exploring some more subtle points of misunderstanding. The remainder of the book moves to issues that are likely to be closer to the interests and experiences of the general reader. In several instances I have taken as my text selected passages from the literature of anthropology and psychology with the intention of highlighting specific points of confusion. The essay roams widely but, I trust, with a coherent theme.

Proximate and ultimate cause and the nature of explanation

Perhaps the most fundamental accomplishment of Charles Darwin was to enlarge for all time the concept of scientific explanation. Virtually every question that one can pose in biology has two very different kinds of answers. Why are leaves green and gentians blue? Why do birds sing? Why do people seek food when they have not eaten for a time? Leaves and flowers are different colors because they contain different pigments. Birds sing and people eat when a combination of environmental conditions stir the nervous system to the generation of more or less predictable behavioral responses. Lengthening days, changing hormone levels, and perhaps the sight of another member of the species conspire to produce song. And with changing chemical signals from the gut and a literally empty stomach, the aroma from the kitchen becomes hard to resist. We can proceed in this vein and describe causes in terms of physiological or biochemical events, but such descriptions will say nothing of the relevant processes that took place in evolutionary time. But we can also assert, with equal validity, that leaves are green because their function is to use the light-harvesting machinery of their chloroplasts to obtain the energy required for growth and reproduction, whereas gentians are a contrasting color so that they are effective in attracting pollinators. Birds sing in order to attract mates and to establish territories large enough to support their offspring during the coming breeding season. Why do we eat? The human genome, no less than that of any other species, packages itself in bodies whose form, function, and behavior serve the successful propagation of the genetic material, DNA—bodies that Richard Dawkins[4] has called "survival machines." Maintaining an energy supply is only one of the more obvious problems that must be solved by every organism, and it should come as no surprise that the human body has self-correcting mechanisms when its energy reserves fall low. Hunger and thirst can be counted on to elicit forms of behavior likely to correct the condition.

The terms *proximate cause* and *ultimate cause* have come into common use to designate these two kinds of explanation. Proximate cause has to do with the characteristics of the organism that one can see—characteristics that are the final expression of the genetic program (the genotype) that is present in the fertilized egg from which the organism grew. Explanations of proximate cause are often couched in the language of physiology and biochemistry and are frequently the subject of experimental manipulation.

Ultimate cause is the province of the evolutionary biologist who is interested in the historical origins of genotypes. Explanations of ultimate cause invoke the concept of adaptation of organisms to their environment as well as evolutionary inferences based on comparative studies of different kinds of organisms. Direct experimental manipulation is not unknown, but it is usually more difficult to achieve. The idea of different kinds of causes can be traced at least to Aristotle, but it was Charles Darwin's contribution to bring ultimate causation to center stage and make evolutionary explanations of "why" a respectable, in fact necessary, part of science.[5,6]

It is essential to understand that explanations of proximate and ultimate cause are not competing means of understanding and it is not necessary to choose between them. Quite the contrary, they involve complementary modes of analysis that generally address very different aspects of biology. With some important qualifications, which we shall consider in Chapter 3, alternatives to Darwinian explanations of ultimate cause are not connected to the rest of science. This is not to deny that for many phenomena there are no single or even convincing evolutionary answers, but rather to assert that an evolutionary framework has proved far and away the most useful, productive, and satisfying way for biologists to approach these questions.

The matter of causation is really not so simply put, and these two major categories can be subdivided. The ultimate cause of bird song, for example, could refer to the evolutionary history that led to singing as well as the purpose to which song is put—its function. George Bernard Shaw suggested that Britain and America are one culture divided by a common language, but the presence of the Atlantic Ocean is not necessary for linguistic confusion. The word "function" as I have just used it is a case in point. Function is also employed by those studying learning in a sense that has no evolutionary implications: For example, the function of the pigeon's pecking the key is to obtain some grain as a reward. In Chapter 3 we shall find some equally confusing ambiguities attached to the word "adaptation."

Proximate cause may have even more dimensions. That kid down the street is in trouble with the police again. What is the cause? Is it the act of buying drugs? Is it the mental state produced by taking drugs? Is it his inability to relate to other people? Is it a deprived childhood? Which of these is correct? Perhaps they are all true. The assignment of proximate cause depends on where one stands in order to view the issue. A pharmacologist, a lawyer, a psychiatrist or psychologist interested in learning and development, and a parent are likely to have rather different things to say about the causes of this behavior. The developmental history of the individual organism may provide as valid a perspective for assigning proximate cause as more immediate and measurable physical or chemical tokens. The difference is not the presence or absence of a physical basis; all developmental and neural phenomena have a physical and chemical basis. The difference, like the larger distinction between proximate and ultimate cause, has more to do with the time frame over which the problem is considered. Nothing of importance in biology can be said to have but a single cause.

The complementary nature of explanations of proximate and ultimate causality has not always been recognized by biologists even in the twentieth century. In fact the evolutionary biologist Ernst Mayr traces much of the failure of communication between geneticists and naturalists prior to about 1936 to the views of major figures like T.H. Morgan, who believed that the new ways of experimental genetics would displace older, more speculative modes of analysis. He writes, in a passage that captures the essence of the distinction,

> Morgan quite clearly did not understand that the clarification of the biochemical mechanism by which the genetic program is translated into the phenotype tells us absolutely nothing about the steps by which natural selection had built up the particular genetic program. It does not tell us why one species has sexual dimorphism and another species does not, nor does it tell us anything about the function of the sexual dimorphism in the life history of the species.[7]

Because so little theory in the behavioral sciences intersects evolutionary theory, there is a similar rift today between biology and significant parts of the social sciences. Specifically, one of the identifiable problems is confusion over just this matter of proximate and ultimate causation. A single example involving humans may be helpful here in introducing the concept. A tendency to avoid brother-sister incest is an example of a widespread human behavior that may well have been molded by natural selection. Discussion of this topic is hardly new. In 1917 Edward Westermarck suggested "that there is an innate aversion to sexual intercourse between persons living very closely together from early youth, and that as such persons are in most cases related by blood, this feeling would naturally display itself in custom and law as a horror of intercourse between near kin."[8] Twenty years later Sigmund Freud[9] rejected this interpretation, partly on the grounds that law and custom should not be necessary to reinforce an "instinct," a line of reasoning that we shall see later on in this book cannot be accepted as a very sophisticated understanding of the determinants of behavior.

The recent evidence for the involvement of evolutionary heritage is (i) the very low incidence of marriage between unrelated Israeli children raised together in kibbutzim from very young ages,[10] and (ii) the high rate of rejection, by the intended couple, of arranged Taiwanese marriages in which the bride-to-be was brought as a young child into the family of the boy and reared as a sibling.[11] These two examples suggest strongly that there are difficulties in establishing adult sexual relationships with a partner one has been closely associated with through childhood. Such a process could be adaptive, as it would tend to prevent the undesirable genetic effects of very close inbreeding. Moreover, for such a system to work in evolutionary time there need be no conscious recognition of who is kin. In the vast majority of cases, only closely related children would be likely to share the necessary intimacy.

The response of one colleague on hearing this argument was "all that proves is that people don't get married if they've known each other too long." With this remark he seemingly intended to shear the phenomenon of any biological interest, leaving it naked for sociologists to examine. The comment is a reasonable (but superficial) statement about proximate cause, phrased, however, as though it were an alternative to an evolutionary hypothesis of ultimate cause. Far from being an alternative, it suggests a plausible mechanism by which the frequency of brother-sister incest might be reduced.

There is in fact evidence from studies of animals that mate choice frequently involves similar but novel individuals.[12] For example, Japanese quail prefer novel first cousins to either familiar or novel but unrelated birds. Corresponding data on humans exist, for spouses chosen freely are more likely to be similar than expected by chance alone, and marriages between first cousins have been common in some societies and at some times.

Insofar as controversy in interpreting this kind of information is based on misunderstanding, it places constraints on a more effective integration of biology, the social sciences, and the intuitive perceptions encountered in philosophy and enduring literature. Let us therefore explore the communications gap in more detail. The uniqueness of humankind is a central tenet of most of the social sciences. In itself this raises no problems, for (at least to a first approximation) every biological species shares a singular set of observable characteristics and can therefore be said to be unique. It is only when the emphasis on uniqueness seems to lift *Homo sapiens* out of the continuum of living forms that blinders go on and tunnel vision ensues. Where this has happened, traditions have been created that are increasingly difficult to maintain in the light of biological findings. First is the supposition that because human behavior is largely learned it is free of genetic influence and even direction. Those who subscribe to this view in its rawest form assert that genetically driven behavior is by contrast rigid and, they say, "instinctual." Second, because the path by which genes produce behavior is difficult to follow, the link between genes and behavior is often, incorrectly, assumed to be weak. Third, although human language is indeed unique, the criteria that define its special character have been narrowed as biologists have uncovered an unsuspected richness in other forms of animal communication. Further, the diversity of human cultures and the range of human choices have traditionally been assumed to be too great to involve any evolutionary steering. This stance, too, has recently been questioned for reasons that are frequently misunderstood. We shall explore these and related matters in some detail, but first we must take a closer look at the current understanding of the evolutionary process.

2

Some Fallacies and Misconceptions

> Alice laughed. "There's no use trying," she said: "one *can't* believe impossible things." "I daresay you haven't had much practice," said the Queen. "When I was your age, I always did it for half-an-hour a day. Why, sometimes I've believed as many as six impossible things before breakfast."
>
> Lewis Carroll, *Through the Looking-Glass*

The recurring national issue of how evolution is (or is not) treated in high school biology texts surfaced once again in California in late 1989 and prompted two revealing letters to the editor published shortly before Christmas in *The New York Times*. The letters are particularly interesting, not so much for what they say, but because of the academic credentials of the authors. One was written by a physician and postdoctoral scientist at an Ivy League university; the second by a professor of sociology at a prominent New York institution. Together they dust off several familiar shibboleths: the theory of evolution does not qualify for classification as fact; evolution deals with history, so it is not subject to investigation by experimentation, and is therefore not proper science; the fossil record fails to support the existence of evolution because it does not reveal every conceivable intermediate form that ever lived; the phrase "survival of the fittest" is a tautology and therefore untestable; evolution above the species level has never been demonstrated in the laboratory; and "not everyone believes in evolution," documented by reference to an "authority" of like mind.

"Evolution is only theory." The word "theory" is used in science to refer to an embracing concept for which there are abundant experimental and observational bases. Cell theory refers to the finding that all known organisms are made up of cells; kinetic theory

refers to the quantitative relationship between heat and the motions of molecules. A theory, in this sense, is a statement about nature that has quite secure underpinnings. It is not subject to formal proof like a theorem in mathematics, and like all scientific knowledge it is of course susceptible to refinement or even replacement by an alternative theory in the light of critical new evidence. The word "theory" is generally not applicable, however, unless it is supported by a body of evidence that makes subsequent abandonment appear very unlikely.

Vernacular usage, however, can be substantially different, ranging from hypothesis—"Your car won't start? My theory is you left the headlights on"—to fanciful—"My theory is that in summer it always rains on Saturdays." A recent examination of high school science textbooks[1] revealed that the word "theory" is widely used in science classrooms synonymously with legend, myth, or any notion that might leap casually to mind, as well as to mean the antithesis of "fact." How much of this misuse in secondary school textbooks is deliberate and intended to keep the critics of evolution at bay may be arguable, but it has the effect of perpetuating confusion on a grand scale.

Biological evolution is both fact and theory.[2] That evolution of living creatures has occurred on earth is substantiated by numerous observations in both biology and geology. Changes that have taken place in the panoply of life can be observed in the fossil record on a time scale measured in millions of years, and studies of natural populations allow us to document smaller changes that have occurred over tens or hundreds of generations. For example, in Britain subsequent to contamination of the environment with soot brought about by the industrial revolution and the massive burning of soft coal, moths (notably *Biston betularia*) showed an increase in the frequency of dark individuals.[3] The pigmentation of the moths is controlled by a gene, and the darker individuals are harder for predators to detect on blackened bark. With efforts to control stack emissions and an improvement in air quality, natural selection has begun to favor lighter individuals once again.

The tools of molecular biology have presented new and independent ways of examining both the results and the process of evolution. For example, proteins consist of sequences of small molecules called amino acids. There are about twenty different kinds of amino acids found in proteins; different kinds of proteins are made of different numbers and sequences of these amino acids. But if we look carefully at one kind of protein, such as the respiratory enzyme cytochrome *c*, obtained from a wide variety of organisms, we find differences in the sequences of amino acids. The differences are not great enough to change the function of the protein—it is still cytochrome *c*—but the differences are larger the more distantly related are the organisms from which the cytochrome *c* was extracted. In fact, one can construct an evolutionary tree based on this molecular information, and the tree conforms closely to the evolutionary tree that is drawn on the basis of the body plans and other characteristics of the organisms themselves.[4] One can therefore see the work of evolution not only in organisms, but in their molecules of protein and nucleic acid,[5] a finding that could not even have been anticipated by Charles Darwin. Evolution is, quite simply, fact.

We also have an understanding about some of the processes by which evolution occurs, and this understanding comprises evolutionary theory. Like all scientific understanding, no one claims that it is complete or perfect. If the evolutionary process were as firmly understood as Newtonian mechanics, it would not be the subject of active research that it is today. In fact, in the next chapter we shall consider some recent refinements to the corpus of evolutionary theory that have particular relevance to the understanding of human nature. But that must wait for a bit.

"The study of evolution is not true science." There is a fallacy that scientific knowledge about the world is obtained only from controlled experiments. For someone who does not wish to believe in the existence of evolution, this is a philosophical position of convenience in which reality is defined away by an arbitrarily restrictive, unrealistic assertion about what constitutes science. Science is about knowing. Much scientific knowledge comes from direct observation, and much understanding comes from comparison and correlation. For example, we distinguish between the flora of rain forests and deserts by simple observation; we derive deeper understanding by correlating observations of plants with climatic and other geographical and ecological factors; and through this process we come to know more about the world.

To know about the world is also to know about its history. Reconstructing historical paths requires evidence from multiple sources, always fragmentary. Specific hypotheses, as in other scientific activity, are often tenuous and subject to revision on the basis of new information. Specific hypotheses, historical or otherwise, are always based on some assumptions, and the assumptions themselves may be questioned and revised. Astronomy, geology, and biology all have historical dimensions, and all are most assuredly part of science. But it would be misleading to suggest that because evolutionary biology is about history, experiments are not possible. Many problems in evolutionary biology, particularly having to do with the evolutionary process, are in fact accessible through experimentation. The most obvious example is a test of selection. The hypothesis that selection for a trait—the differential reproduction of those individuals possessing a heritable trait—can increase the frequency of occurrence of the trait in the population of progeny has been put to experimental test and validated in many instances. Suppose, however, that in a particular instance the experimental test failed. Since the hypothesis includes the assumption that the trait is heritable, a possible (even likely) explanation for the negative outcome of the experiment would be that in this particular case the trait is not in fact heritable.

"The fossil record does not support the idea that evolution has occurred." Here the ground shifts under our feet. The denial of history as science is replaced by criticism of the actual historical record. Because the known fossils have not revealed every conceivable evolutionary intermediate, the argument goes, there must be something wrong with the concept of evolutionary change.

Analogy with social history may clarify the flaw in this reasoning. The existence of the Roman Empire is accepted long after all witnesses have ceased to exist. The acceptance of historical events in culture therefore does not depend finally on living memories but on various kinds of records—written, architectural remains, artistic creations, and ultimately, as the mists of time grow thicker, a handful of miscellaneous artifacts created by human hands. The depth of our understanding of the various periods in history varies with the richness of the record, but we never question the existence of history due to the record's being incomplete.

Not knowing how a Caesar spent some particular month of his reign does not alter our conviction that he and the empire existed. But we meet the equivalent of this argument in critics of evolution who assert that because every conceivable intermediate in every conceivable phyletic line did not have the good fortune to become fossilized (for many kinds of organisms an event of intrinsically low probability) and the fossils have not subsequently been discovered (also an improbable event), evolution could not have occurred. The reality, however, is that a century of post–Darwinian paleontology has enriched our understanding of evolutionary history and will continue to do so in the future. The existence of Precambrian fossils older than 600 million years was unknown in Darwin's time and remained unknown until scarcely more than thirty years ago. Cellular forms of life are now known from the earliest sedimentary rocks that have survived the rigors of metamorphosis, dating from at least 2.7 billion years ago, and the first nucleated cells appear about 1.4 billion years ago.[6] Similarly, the great explosion of invertebrate body plans of the Cambrian period, whose fossils were locked away in the Burgess shales of the Canadian Rockies, has come to be understood only in the last twenty years.[7] Much of our fragmentary understanding of human evolution is derived from fossil material discovered in East Africa in the lifetimes of many of us. Paleontology continues to this day to add to the catalog; the first fossil whale with rudimentary hind limbs has been reported as I write these words.[8]

"Evolution is not testable." Here the argument runs something like this: Evolutionary theory is adequately summarized by the phrase "survival of the fittest." But if we measure fitness by those who survive, evolutionary theory is nothing but a tautology. Of course the fittest survive; there is no alternative!

The fallacy is in supposing that the concept of natural selection is adequately embodied in Herbert Spencer's oft-quoted aphorism "survival of the fittest." There are, moreover, two distinct problems created by this catchy phrase.

To make the issue clearer, let us look first at an economic analogy. Large numbers of savings and loan associations have failed during the last several years. One might describe the pruning of the industry as "survival of the fittest," suggesting that those institutions that remained solvent had, in some unspecified way, been more successful. The phrase has suggestive implications, but it hardly forms an economic theory with any detailed explanatory power. One needs to know about internal factors such as quality of management and its strategy of investment, the nature of the banks' assets, external factors such as local markets and the impact of regulatory rules, as well as how these factors interact in particular cases. No one expects a theoretical construct of any interest to emerge without such detailed analysis. In this respect, evolution by natural selection is no different. I shall offer some examples presently.

The second problem has to do with the word survival. In evolution, the issue is not so much survival as it is reproductive success. Natural selection is a kind of sieve, in which those organisms that reproduce most effectively are the ones whose genes are represented most abundantly in succeeding generations. An organism can therefore survive to a ripe old age without reproducing, and if it does, it contributes nothing to future evolution. This is a matter to which we will return later, but let us continue here with the charge of tautology. A critic would perhaps argue that substituting "reproduction" for "survival" does not change the circularity of Spencer's characterization of evolution.

Evolutionary change indeed results from differential reproduction of individuals with different complements of genes (i.e., different genotypes), but as with the banking analogy one must take the next step and inquire why one genotype is more successful than another. Specifically, what characteristics of the organism (i.e., which phenotypes) enhance the reproductive success of one genotype relative to another? And what is the process by which this comes about? For example, does one breed earlier, or lay more eggs, or take better care of its young, or make more pollen, or exploit resources more efficiently, or tolerate hard winters better? Does one genotype have the edge only in particular environmental circumstances? And if so, why? The list of possibilities is in principle long, but intimate knowledge of the organism often provides a clue.

These kinds of questions can frequently be answered by comparative observation, or more directly, by experiment, but only when such issues are addressed with specific hypotheses can one begin to apply the concept of adaptation by natural selection to arrive at an understanding of what in the organism's life-style contributes to its "fitness." To assert that evolutionary theory is simply "survival of the fittest," full stop, end of lesson, is to reduce a rich theoretical construct with great heuristic value to a mere caricature of itself. Doing so hardly does justice either to Darwin's considerable insight or to the kinds of issues that are in fact addressed by contemporary evolutionary biologists.

"Evolution beyond the species level has never been observed in the laboratory." It is not clear why anyone should expect this to have occurred. Evolutionary change of this magnitude requires more time than is available in a human career. Evolutionary changes of smaller magnitude, however, have been observed in natural populations and have been produced experimentally numerous times.

"If evolution has occurred, contemporary species that are members of older lineages (for example, reptiles) should be 'intermediates' in the evolution of species from more recent lineages (for example, mammals)." This is a curious argument, for clearly an organism living today cannot have been an ancestor of an organism that existed hundreds of millions of years ago. Moreover, contemporary reptiles have had evolutionary histories of their own during the entire time that mammals have existed. Although the defining characters of taxonomic groups have changed more dramatically in some lineages than in others (permitting us, for example, to distinguish mammals from reptiles), no species living today can be considered ancestral to another. This would seem obvious, but let us look at a couple of arguments that have taken root in the soil of this misconception. I draw them both from Michael Denton's *Evolution: A Theory in Crisis.*[9]

Denton makes an effort to set the reader's place at the table, so each of these examples requires a few more words of background. Taxonomists are biologists who describe and classify organisms. Without their doubting the reality of evolution, taxonomists nevertheless have a variety of schemes for constructing hierarchical relationships between organisms. There are several reasons for this flexibility. First, the criteria for distinguishing taxonomic groups above the level of species (e.g., genera, families, orders) are quite arbitrary and vary from one kind of organism to another. The reasons for this should be obvious: One would be hard-pressed, for example, to classify different kinds of automobiles by the same criteria that would be useful in classifying airplanes. More important for our immediate argument, there are different schools of thought as to how a classification scheme should relate to evolutionary history. One school constructs hierarchical relationships on the basis of overall similarity; the more characters two organisms share, the closer they are placed in the scheme of classification. Another school (cladists) distinguishes between ancestral and derived characters and constructs classifications (in the form of branching trees called cladograms) that are based only on the latter. Although both approaches lead to tree-like hierarchies (an example of which was mentioned above in the discussion of cytochrome *c*), the results of such classifications can be different. For example, the first school places crocodiles among the reptiles, a seemingly intuitively obvious scheme even to the most casual observer of living things. But because crocodiles arose from a lineage that also includes birds, and after the branches leading to lizards and turtles, cladists assert that the traditional Class Reptilia is an artificial construct that does not reflect proper evolutionary relationships. It is important to recognize that these issues do not reflect disagreements about the *existence* of evolution and they do not address the mechanisms of the evolutionary *process*. They have to do with the practice of classification and the details of evolutionary history.

Cladists attempt to construct their trees from analyses of characters and then draw conclusions about evolution. Denton, however, has ignored the latter part of this intellectual process, likening branched cladograms to hierarchic typologies that were drawn in the ninteenth century. He then proceeds in the following vein:

> The trees of the typolgists . . . merely represented the abstract logic which underlies all hierarchic systems of relationships, the branches and the interconnecting nodes being purely theoretical. The fact that all the individual species must be stationed at the extreme periphery of such logic trees merely emphasized the fact that the order of nature betrays no hint of natural evolutionary sequential arrangements, revealing species to be related as sisters or cousins but *never* as ancestors and descendants as is required by evolution. The form of the tree makes explicit the pre-evolutionary view that it is discontinuity and the absence of sequence which is the most characteristic feature of the order of nature. [pg 132]

This assertion is quaintly confused on two counts. The first is expectation. Contrary to Denton, that contemporary organisms should seem to fall into nested hierarchies of relatedness is exactly what one would *expect* of evolution. There is no reason whatsoever to predict a continuous gradation between all living forms. The expectation is a straw man, for the very act of speciation produces discontinuities that are amplified by further evolutionary change.

The second misleading point has to do with the processes by which contemporary evolutionary biologists construct such hierarchical trees (cladograms). This kind of analysis is useful primarily *because* it makes explicit likely evolutionary relationships. Furthermore, as mentioned above, the mathematical techniques can be applied to features of entire organisms or sequences of amino acids in single proteins, depending on whether the object of study is a group of related species or a family of similar (in fact homologous) proteins. The mathematical algorithms are designed to find the most parsimonious trees—that is, the trees with the shortest branches—on the reasonable premise that the most likely evolutionary relationships will have involved the fewest number of evolutionary changes. With a small number of character states (the general term for data used in constructing trees), the analysis not only reveals the most likely sets of relationships, but it also predicts the probable ancestral conditions that existed at the branch points in the tree. But without further assumptions (which cladists avoid making), this kind of analysis says nothing about the times at which these ancestral states existed, and it gives no clue as to what other characteristics the ancestral species possessed.

To say it again, the species that lie at the ends of the branches of these hierarchical trees are different from any of the ancestral forms from which they arose. They are indeed in a sense "sisters" and "cousins" of one another. This mode of analysis furthermore predicts that ancestral forms probably had particular combinations of the characters under consideration, but it does not tell us what those ancestral organisms actually looked like in their totality.

"Evolutionary change is the result of blind chance." As has been argued elsewhere,[10] just the opposite is true. There is a large element of chance in the processes of mutation, extinction, and the genetic changes that take place in small populations. But the process of natural selection is, on average, a process of sifting in which those combinations of genes that produce organisms best able to reproduce successfully in the environment in which they find themselves are the genes and gene combinations that make their way into the future. The sifting is the very antithesis of chance.

It is hard to imagine evolution producing such complicated results. This objection is sometimes coupled to the previous one on the assumption that evolution is the result of a large number of totally independent and individually improbable events. But it also illustrates how the human brain does not accommodate readily to scales of space and time that lie far outside the realm of personal experience. For example, individuals are frequently willing to judge evolutionary hypotheses by an exercise of the imagination or the test of whether they conform to common sense. In fact, the willingness to engage in this exercise seems totally unrelated to the individual's actual knowledge of evolutionary theory.

Let us start with a trivial example. In *The Blind Watchmaker*,[11] Richard Dawkins coined the phrase "The Argument from Personal Incredulity" to describe the Bishop of Birmingham's inability to imagine why natural selection might have made polar bears white. A more serious example is Michael Denton's[12] rejection of macro-evolution on the basis of estimates of probability of the fallacious nature described above. His argument is basically that of Personal Incredulity, rhetorically capped with the words of Lewis Carroll that are cited at the opening of this chapter.

But let us look at some other examples of biological phenomena where imagination is challenged. Seventy-five years ago knowledge of the molecular architecture of cells was essentially nonexistent. One could look down the tube of a microscope at the curdled remains of a cell that had been pickled in formaldehyde and stained with one or another dye to make the nucleus visible and wonder, as did the distinguished geneticist William Bateson, just what it could possibly mean that the genes, whose effects were evident in breeding experiments, seemed to be arranged in linear sequence on chromosomes:

> The supposition that particles of chromatin [the withered remains of the nucleus visible in the light microscope], indistinguishable from each other and indeed almost homogeneous under any known test, can by their material nature confer all the properties of life surpasses the range of even the most convinced materialism.[13]

Bateson's riddle was understandable, and of course it persisted until James Watson and Francis Crick discovered the two-stranded, helical structure of the DNA molecule in 1952 and until the genetic code was subsequently shown to reside in the precise order in which the four kinds of constituent molecular subunits (the nucleotide bases) are strung together to make the genetic material

DNA. What had been simply incomprehensible in the early decades of this century has now become crystal clear and in principle within the grasp of understanding of anyone who knows only a modest amount of elementary chemistry. The world of molecular shape and structure is still being revealed by technological advances, without which imagination of the very tiny is a poor tool.

Consider now a single cell containing that information stored in the molecular structure of its DNA. Suppose further that the cell is a fertilized egg, destined to grow into a sparrow or a sunflower. Unlike the example of the physical basis of the genetic code, the development from egg to adult organism is a process that occurs over time. Although we know quite a bit about the way the information in the DNA is decoded to make proteins, allowing the cell to metabolize and live, we cannot (yet) say too much about how development happens: How the cell proliferates by successive divisions and how the progeny specialize to form skin, bones, liver, brain, or, as the case may be, leaves, roots, and flowers; and how the various specialized cells manage to cooperate to produce a functioning organism. We will have more to say about development later, but for the moment let me suggest that it is only our familiarity with the results of these events that leads us to accept them as real. If development were a rare event, observed by only a few people in their lifetimes and understood no better than it is today, many might harbor suspicions that it is too complicated for science to explain, or perhaps even to happen. But we commonly see organisms develop on our scale of time, and we accept the process as part of nature. Familiarity is the handmaiden of common sense.

We do not see evolution happening every day. We see only the results, and so unless we make an additional effort to inform ourselves, our common sense, honed as it is by personal experience, gives us little practical guidance.

3

Evolutionary Theory Since Darwin

No man, when he hath lighted a candle, covereth it with a
vessel, or putteth it under a bed; but setteth it on a candlestick,
that they which enter in may see the light.

For nothing is secret, that shall not be made manifest; neither
any thing hid, that shall not be known and come abroad.

Luke 8:17-18

Our understanding of evolution is built on a Darwinian tripod of
three fundamental observations: (1) living organisms are units of
organization that reproduce; (2) individuals differ from one
another, and some of their differences are inherited, and (3) in a
population, individuals enjoy differing degrees of reproductive success based at least in part on these heritable differences. This differential reproduction is what is meant by natural selection.

Darwin's intellectual triumph was to see that natural selection is
the principal cause of evolutionary change, a perception that
remains at the core of evolutionary theory after more than a century
of subsequent investigation. This is a rather solid accomplishment
by any reckoning, but it is particularly remarkable given Darwin's
total ignorance of the rules of genetics and the basis for heritable
variations. But in Darwin's day, no one save the monk Gregor
Mendel had any experimental evidence about how inheritance
worked, and the world did not become aware of Mendel's observations on the genetics of peas until some years later. The formal
rules of genetics, including the recognition that genes—entities
controlling the appearance of discrete characters in the adult organism—are strung out in linear sequences in chromosomes, were
worked out in the opening decades of the twentieth century.

In fact, Darwin's struggles with this aspect of the problem led
him to entertain Lamarckian ideas about the possibility that char-

acteristics acquired in an organism's lifetime might be inherited by its offspring. The inheritance of acquired characteristics is actually rendered untenable by the irreversible nature of the reading of the genetic code. Aside from mutations in the code itself, there is no way for novel changes in the parts of cells or organisms that occur during the lifetime of the organism to find their way back into the DNA (deoxyribonucleic acid, about which more below) that makes up the genetic code. This is an understanding acquired during the last forty years, so it clearly could not have influenced Darwin's thinking. Historically, however, by the turn of the century the inheritance of acquired characteristics had begun to look unlikely for a somewhat different reason. At the end of the ninteenth century August Weismann recognized that early in development a few embryonic cells become committed to a future as eggs or sperm (the germ cells) and the remainder become dedicated to supporting roles as body tissues (somatic cells). This sequestering of the germ line as undifferentiated cells seemed to buffer it from changes acquired by the organism during its lifetime. The sequestration of the germ line occurs in the so-called higher animals, but in some organisms somatic cell lineages maintain a reproductive potential, and for these organisms the inheritance of acquired characteristics remained a theoretical possibility until the molecular biology of the gene began to be understood.

In an important respect, however, the separation of germ and somatic tissue early in development does buffer future generations from one class of environmentally induced changes—mutations that occur in the DNA itself. If a mutation should occur in the DNA of a somatic cell, for example from liver or muscle, this modified gene forgoes any possibility of passing to future generations and ceases to exist when the organism dies.

Curiously, it was not until the late 1930s and 1940s that evolutionary biology and genetics were effectively melded. This accomplishment, often referred to as the "modern synthesis," provided a broader theoretical basis for studying evolution and has guided much research in population biology for the last fifty years.

Natural variation and its sources

With trivial exceptions, no two sexually reproducing organisms are identical. Unlike molecules of one kind of chemical, once you have seen one individual of a particular plant or animal species, you have not seen them all. This vast amount of variation is the grist for evolution's mill. But understanding patterns and processes of evolutionary change requires attention to variation at two distinct levels: genotype and phenotype. Genotype means the genetic endowment of an individual. For sexually reproducing organisms with two sets of chromosomes, like ourselves, genotype is determined when the sperm and egg join to form a zygote. Although the word genotype is often used in this way to refer to the total genetic endowment—the genes at all loci—it can also refer to the particular forms of a gene (the alleles) at a single locus, for example, the human genes controlling eye color or sickle-cell anemia. The phenotype of an individual is the constellation of traits that we see when we look at the organism and may include morphological, behavioral, physiological, and biochemical features.

In a fundamental sense, evolution is the sifting of genotypes; however, differential survival and reproduction occur among phenotypes. In the world of interacting organisms, phenotypes are the agents of the genotypes, and it is the phenotypes—the organisms themselves—that compete and whose performance determines reproductive success. Obviously, natural selection can act only if the basis for differential survival and reproduction of phenotypes is heritable. Phenotypic differences that are not the result of underlying genotypic differences therefore cannot serve as the basis for evolutionary change. It is for this central reason that we must be concerned with how the genotype becomes translated into phenotype and with the effects of the environment on this process. This relationship will receive our full attention in Chapter 5.

The question of how much genetic variation actually exists in natural populations is critical to evolutionary theory, but it is one that has been effectively addressed only in the last generation. Observant naturalists have been aware of phenotypic variation for a much longer time, and breeders of both plants and animals know that artificial selection can be very effective in producing heritable change. Neither of these observations, however, provides a quantitative measure of the underlying genetic variation. Starting about twenty-five years ago techniques became readily available for spotting variation in the speed with which soluble proteins extracted from plants and animals migrate in an electric field. Because in general a single gene codes for a single protein, variation in the physical properties of protein molecules of the same kind (e.g., that perform the same enzymatic function) is a convenient indication of variation in the gene that codes for the protein. In the following paragraphs we shall see in more detail the nature of the code and its relation to the structure of a gene. The immediate point, however, is that there is considerable genetic variation in natural populations. Based on data from a few soluble proteins and species that have been examined extensively, 15–59% of genetic loci exhibit gene polymorphism (i.e., more than one version of the gene exists in the population), and 3–15% of genetic loci in each individual organism are heterozygous, which means that there is a different version of the gene (a different allele) at this locus on each of the two parental chromosomes.[1] Some of this variation seems to have little or no selective advantage in the current environment, whereas some may code for slightly different functional forms of the protein. It is the slow accumulation of alleles coding for slight differences but with little or no selective advantage that allows us to construct evolutionary lineages for individual kinds of protein molecules, as described in the previous chapter. The longer the time of evolutionary divergence, the greater the accumulation of these selectively inconsequential changes in the gene.

Mutations are heritable alterations in the genetic material, which in most organisms is DNA (deoxyribonucleic acid). DNA is a long helical molecule consisting of two twisted strands, each a string of nucleotide bases of four kinds (adenine, guanine, cytosine, and thymine, usually referred to as A, G, C, and T). Genetic information is coded in the sequence in which the bases are strung together. Specifically, DNA codes the order in which amino acids are connected together during the synthesis of individual proteins; a particular triplet of bases signifies that a specific amino acid will be added to the growing protein. The sequence of nucleotides in a gene thus determines the sequence of amino acids in a protein much as the sequence of dots and dashes in Morse code specifies the sequence of letters in a message.

The key to understanding how a dividing cell partitions the coding material equally to its daughter progeny is contained in the helical structure of the DNA molecule. The two strands of the helix are complementary, in that a given base on one strand always pairs (through weak chemical forces) with a particular one of the other three: A pairs with T and G with C. This means that the base sequence on one strand determines the sequence on the other. When a cell divides, every helix uncoils, and each strand serves as a template for the synthesis of the complementary strand. The result is two identical copies of the double helix, one for each daughter cell.

Various kinds of mutations can occur, but one of central importance is the substitution of one of the four nucleotide bases for another while DNA is being copied in preparation for cell division. This brings about a small change in the genetic code in one of the daughter cells, which in turn can alter one of the amino acids in a protein. If the mutation occurs in the germ line (ova or sperm), the mutation can be passed on to subsequent generations.

Mutations can also involve insertion or deletion of long sequences of the code, or they can arise from rearrangements of existing sequences. The magnitude of the genetic change caused by a mutation is not necessarily correlated with the magnitude of its phenotypic consequences. Changing a single base may have no phenotypic effect at the level of the organism, or it may render a critical protein dysfunctional and lead to the organism's death. For instance, the change of a single base, leading to the substitution of a different amino acid in the blood protein hemoglobin, is responsible for the condition in humans known as sickle-cell anemia.

During cell division genes are packaged in bundles that can be seen in the microscope. Because these structures absorb certain kinds of stains, they were given the name chromosome (which simply means colored body) in the early days of microscopy, well before their relation to the material of the genetic code was understood. They represent the existence of groups of genes that are linked together. But because in the process of their duplication pieces of chromosome can exchange (a phenomenon known as crossing over), the linkage groups are not stable and are subject to recombination. Recombinants of multi-locus linkage groups arising in the germ line may lead to distinct phenotypes in the next generation. Recombination is therefore an important source of genetic variation, and it may be one of the major reasons for the evolution of sex.

The evolution of sex actually poses an enigma. Asexual organisms will produce offspring that are uniformly like the parent, which may be fine if the parent is well suited to its environment and the environment is unchanging. But in environments that are spatially or temporally heterogeneous, it may well pay to produce offspring with considerable phenotypic diversity, hence the presumed advantage of sexual reproduction. The flip side of this, however, is that recombination also serves to break up favorable combinations of genes. Sometimes pieces of a chromosome become inverted, and this suppresses crossing over and recombination. Inversions may therefore serve to hold together particularly favorable complexes of genes.

Understanding the organization of the coding material in organisms with nucleated cells (eukaryotic organisms) has been profoundly expanded by recent discoveries in molecular genetics.[2] Many genes consist of coding regions (exons) interrupted by intervening sequences of DNA (introns) that have no coding function. In fact, it is not known whether introns have any function, but their existence raises the possibility that many proteins may be put together in modular fashion. Thus the evolution of a protein with a new function might not necessarily involve gradual changes in the base sequence of a particular gene but could arise by putting together exons from already existing proteins, a process which is sometimes referred to as "exon shuffling."

It is also now clear that a significant fraction of eukaryotic genomes consists of repetitive sequences. Some are repeated many thousands of times; others occur in just a few copies. Many abundant proteins such a globins and actins are coded by small families of similar genes, presumably derived in evolutionary time by duplication and subsequent divergence. In many cases, however, the members of these families are now dispersed throughout the genome, occurring on different chromosomes. Not only have the genes diverged in base sequence, but they have also come under different regulatory control, allowing them to produce not only different gene products but to be active in different tissues or at different times during the development of the organism from a fertilized egg.

A recent remarkable discovery is that many repetitive DNA sequences are able to move around in the genome. Transposition of sequences was deduced a number of years ago by Barbara Mc-Clintock from her painstaking genetic and cytogenetic studies of maize, but her findings were initially hard to appreciate given the understanding of molecular genetics that existed at the time. Modern DNA technology has not only confirmed her observations, but has allowed a much more detailed analysis of transposition. Many of the transposable elements duplicate and move at the same time, with one copy remaining at the original location.

The existence of this duplicative transposition, together with the observation that most organisms appear to possess far more DNA than they seem to need to code for all known functions, has led to the notion of "selfish DNA."[3] If phenotypes are simply the genes' way of making more genes (a contemporary recasting of the observation that a hen is the egg's way of making another egg), and if certain sequences of DNA do not make any contribution to the phenotype, perhaps these sequences simply replicate themselves within the environment of the rest of the genome.

Forces of evolutionary change

Evolution is change over time in the genetic composition of natural populations. What factors are responsible for these changes? The conventional approach to this problem is to extrapolate from microevolutionary processes, which can be observed either in the laboratory or in natural populations. In this view, which is held by many evolutionary biologists, profound macroevolutionary changes have arisen as a result of an accumulation of smaller, observable microevolutionary changes. Something may well be missing from this part of the conceptual scheme, which is one reason why evolutionary biologists are yearning for a more detailed knowledge of the genetic control of development and a deeper understanding of how the genomes of eukaryotic organisms are organized.

What are the identified forces at work in microevolution? Four factors clearly influence the frequencies of genes and contribute to evolutionary change: mutation, gene flow, random drift, and selection.[4]

Mutation. Although mutation is clearly the ultimate source of variation, spontaneous mutation rates are so low that mutation alone is not an important cause of changes in the frequencies of genes in natural populations. Mutation rates are generally of the order of 10^{-4} to 10^{-6} per gene locus per generation, so it would take millions of generations for mutation, working by itself, to shift the frequency of any particular gene in a population.

Gene flow refers to the migration of genes between populations due to the movement of organisms in which they reside. Although there are few reliable estimates of the magnitude of the exchange of genes between natural populations, a major consequence of any exchange should be to counteract the tendencies for populations to diverge. As an aside, the present view of how new species form involves the spatial separation of populations and the absence of effective gene flow for a sufficient time that the populations begin to evolve independently of one another. If sufficient time passes and enough evolutionary change occurs, the populations may then prove incapable of interbreeding even if the geographical barrier that originally separated them should cease to exist. At this juncture they may have become genetically incompatible, or they may be kept from interbreeding by differences in the timing of reproduction, different requirements for habitat, differences in behavior, or any one of a number of other so-called *isolating mechanisms*.

Random drift. Chance enters evolution in a number of ways, of which mutation is only one. Consider how some human family lineages expand and others die out; disease, accident, war, biased sex ratios at birth, and infertility of individuals all can contribute. In natural populations of organisms, environmental fluctuations, which are to some extent random, make the success of any particular lineage unpredictable. An obvious if dramatic example of such a chance event with major consequences is the postulated asteroid impact invoked to explain the large-scale extinctions and the demise of the dinosaurs at the Cretaceous-Tertiary boundary some 65 million years ago.

A second type of chance event is the sampling error that occurs each generation when parental genes are passed on to offspring. In the absence of mutation, gene flow, or selection, one might predict that gene frequencies would remain the same from one generation to the next, but this will be true only if populations are very large. The probability is very close to 0.5 that each human child will be a son, but reflect on the number of families that have only daughters or only sons. Similarly, in small populations of interbreeding individuals, chance deviations in the distributions of genes in fertilized eggs will occur in each generation. This phenomenon can produce significant shifts in the frequencies of genes and draws particular importance from the suspicion that many speciation events begin in small, relatively isolated populations.

Selection is far and away the most important microevolutionary force. Selection is the differential reproduction of individuals with particular heritable traits and can produce shifts in the frequency of a gene in a population many times faster than mutation. Moreover, as we have previously observed, selection is not a random or chance process. Selection *is* the more successful reproduction by particular individuals *because* of the heritable characteristics they possess. Much of the rest of this book has to do with factors that contribute to reproductive success and natural selection.

The sometimes elusive concepts of heritability, adaptation, and fitness

These words are used freely by biologists and non-biologists, usually without precise definition and frequently ambiguously. So a bit of effort exploring these thickets now should make our subsequent journey easier.

Heritable and heritability; the pitfall of population thinking. In common usage, heritable means passed from parents to offspring, as with a heritable title in English aristocracy; in a biological context, a genetic basis is causally implied. For example, natural selection requires that individuals in a population not only vary, but that at least some of the variation reflect the presence of different genes and thus be *heritable*.

The term *heritability*, however, has an exact statistical definition that makes it trickier to use unambiguously. This is a matter of more than casual interest, because it provides a window on a more substantive issue. Imagine a population of individuals that vary with regard to some physical trait. One of the analytical tools of population geneticists is to try to correlate the variation with particular factors including aspects of the environment. Sometimes it is possible to partition the variation (technically, the variance) into genetic and environmental components, and if several influences are acting independently of each other, the sum of the individual variances will equal the total variance. *Heritability* is the fraction of the total variance that can be ascribed to genetic factors and is a number that varies between 0 and 1.

Let us consider an example. Suppose we were to measure the skin color of a number of Swedes living in Stockholm. We would find some variation, but most of it would probably have to do with sun lamps and Mediterranean vacations, and the heritability would be close to zero. On the other hand, suppose our population consisted of employees of the United Nations. In this case we would surely find that skin color has a higher heritability. So the first point about the heritability of a trait is that it depends on the population in which it is measured. It is not a number that has any meaning for other populations, or even for other environments. Its value depends wholly on the context in which it was measured.

To return to our Stockholm population, suppose that all of the individual members have blue eyes. There is no variation in eye color, so (in this population) heritability of blue eyes is, by definition, zero. But we know that eye color is under simple genetic control—it is heritable. This illustrates the second point about the term heritability: Heritability clearly does not mean the degree to which a trait is genetically determined. Consequently, a measure of heritability says exactly nothing about why any individual does or does not possess the trait. It does not speak to the role genes play in controlling the expression of the trait. Failure to understand these distinctions has a way of perpetuating the nature-nurture morass, to which we shall return later.

Heritability is a measure of genetic *variation* (for that trait) *in a particular population*. The greater the heritability, the greater the genetic variation. A nonzero value of heritability is necessary for natural selection to effect a change in the character, but low (even zero) heritability may mean that natural selection has been intense.

What do we mean by "adaptation"? Adaptation sometimes means a state of being and sometimes it means a process. The modified forelimb of a seal—its flipper—is an adaptation for swimming, but the term adaptation can also be used to describe the gradual process of modification through evolutionary time by which the flipper was produced by natural selection. Some have been willing to characterize as an adaptation any characteristic that contributes to survival or reproductive success. Others, more circumspect, would reserve the word "adaptation" for features that have clearly been molded by natural selection for the functions they now serve.[5] The distinction is not trivial, for it forces us to contemplate how evolution actually occurs. For example, Darwin observed that the sutures in the skulls of young mammals had been suggested by others to be "a beautiful adaptation for aiding parturition," but as sutures are also present in the skulls of birds and reptiles, he saw that their presence in mammals must be a more general property of vertebrates, however useful or even necessary sutures may be for mammalian birth. An adaptation, in the more restricted view, is therefore a result of a particular kind of historical process: natural selection.

How should we characterize the features of an organism like the sutures of the mammalian skull that did not arise in their present role by natural selection but nevertheless function to promote survival and reproduction? Sometimes organisms find themselves in new environments and are able to employ existing structures or behaviors for entirely new ends. Such features have traditionally been called preadaptations, but the term "exaptation" has recently been proposed on the logical grounds that if an adaptation has been shaped by natural selection, there can be no such thing as a preadaptation.[6]

This terminological issue and the creation of discrete categories of "aptations" must not distract us from the underlying complexity of the process of adaptation. There are not simply two kinds of histories for characters, those molded by natural selection for current function and those evolved for other purposes and then co-opted later. Every character experiences a diversity of selective environments and is subject to an array of constraints. Whether it is properly called an adaptation is therefore a function of how long it has been subject to the current selective regime, as well as how we choose to define the character. An adaptation only makes sense when referred to a particular environmental problem the organism must "solve." As time passes, selective regimes change, and a particular character may cease to be the adaptation that it was in a previous environment. It may then be an exaptation in its new environment, and as the character evolves further in response to new selection pressures, it again becomes an adaptation. Throughout this process the character itself is changing, so that after some period of evolution we may fail to associate it with the original character and therefore give it a new name.

Limits to adaptation. As we saw above, evidence for evolutionary change over time is frequently linked with natural selection as the mechanism of that change. As a consequence, there has been a tendency to view all traits of organisms, living and fossil, as adaptations. This tendency is reinforced by analyses of function that apply ideas from engineering design. Organisms are frequently discussed as though they consisted of a number of independently engineered solutions to a series of discrete problems posed by the environment.

When pressed, however, most biologists recognize that organisms are integrated wholes, and that there are necessary trade-offs in the evolutionary process. A feature may not represent the optimal solution to one particular problem because of selective constraints imposed by other aspects of the environment. For example, in natural populations of guppies (a fish), sexual selection has led to brightly colored males in areas where predators are absent.[7] In the portions of streams where predatory fish are present, male guppies are much more cryptic. Male coloration therefore appears to represent a balance between two opposing selective forces. In any particular population, male coloration may not represent an optimal solution either for attracting mates or avoiding predators. Phenotypes are necessarily compromises.

Another reason that evolutionary solutions may not be optimal is that there are frequently multiple solutions to particular problems. Adaptive solutions are frequently compared with a landscape of peaks and valleys in which the highest points represent solutions. Evolution moves phenotypes to the peaks, but which peaks? Not necessarily the highest (i.e., the best solutions), but generally to the nearest. The chambered nautilus has a large but lensless eye that peers at the world through a pinhole—clearly not the best way to build an eye, but seemingly adequate to the needs of this mollusk.[8] Pinhole optics represent a foothill on the landscape of all known adaptive peaks that might be occupied by eyes. The historical evolutionary path of a lineage may therefore determine which peaks are climbed, viz., which adaptive solutions are in fact found. A corollary of this historical contingency is that evolution is largely irreversible. There is no turning back, and as a result, as environments change, some lineages are able to adapt, whereas others fail and die out. Extinctions have been common in the history of life.

There appear to be a number of other significant constraints on natural selection. Some arise from chance. For example, random drift can, in principle, lead to differentiation of populations and fixation of alleles, even in the face of opposing selection pressures. Second, single genes may have effects on many aspects of the phenotype (a phenomenon called pleiotropy). For instance, defects in the metabolism of particular amino acids may influence phenotypic characters as apparently diverse as body color and behavior. We shall have much more to say about behavior as a phenotypic character later on, but a reason why a perturbation of amino acid metabolism can disturb both pigmentation and behavior is that amino acid derivatives are involved both in the formation of pigments such as melanin and the mechanisms by which many nerve cells communicate with one another.

Selection on one aspect of the phenotype may therefore cause changes in another feature of the organism, even though the latter is not itself the object of selection. Selection for a rapid rate of development might produce individuals of small size, not because there is any selective advantage to small size per se. Small size might simply be a consequence of a rapid rate of development. Similarly, different body parts may be constrained to grow in certain ways or in certain proportion to other body parts because of the mechanics of some other feature of development.[9]

Determining what is an adaptation can be compromised by our inability to explore the evolutionary path that led to the present. This is of course a problem common to all questions of history. Hypotheses about adaptation are not readily falsified. How "unadapted" is the condition with which we might compare? What was the starting point for the evolutionary history of the trait of concern? Comparative data from close relatives of the species can help enormously in unraveling these puzzles. Examination of similar species (or even other populations of the same species) can help resolve the nature of the ancestral condition and the nature of the changes that have occurred to separate these two forms. For example, if particular features of the life history of an organism are hypothesized to be adaptations that enable it to live in arid conditions, we may be able to reject the hypothesis if we find that close relatives occupying more lush environments also possess these traits. Conversely, if several independent lineages that occupy similar environments possess similar life-history characteristics, and if we can demonstrate that the likely ancestral condition in each case was different, then a good case can be made that the traits are indeed adaptations and the result of convergent evolution.

If we recognize that the current relationships of phenotypes to environments are the result of selection's interacting with other agents of change and with a number of constraints—historical and developmental—evolution appears to be a much more complex and intricate process than were natural selection alone to determine evolutionary outcomes.

The slippery meaning of "fitness". This word frequently must be understood intuitively, as in "survival of the fittest." An organism is fit by virtue of those features that contribute to its reproduction. Survival may also be important, but only to the extent that it contributes to successful reproduction. The white coat of the polar bear contributes to its fitness by making the bear less likely to be detected as it stalks a seal for dinner. The capacity of a kangaroo rat to form concentrated urine contributes to its fitness for life in the desert. These features of polar bears and kangaroo rats are adaptations to meet particular environmental challenges.

Sometimes it is possible to characterize the fitness of a biological structure in quantitative terms; for example, the optical performance of eyes can be assessed against theoretical physical designs.[10] Usually, however, this is not a particularly useful exercise because natural selection does not have the luxury enjoyed by an engineer, who may choose from an array of materials and assemble them in any way that is consistent with the solution of the problem. Evolving organisms must cope with an evolutionary challenge by utilizing their genetic heritage and any small mutational changes that may come their way. Their genetic heritage may limit their use of materials, or it may constrain how those materials are assembled. In a number of respects, the vertebrate eye comes close to optical perfection, but no animal has constructed a lens that is corrected for chromatic aberration, a problem readily solved in the manufacture of cameras and microscopes. Similarly, the vertebrate retina is put in the eye backwards, so that light must pass through several layers of neurons before it reaches the rods and cones. This latter quirk is an evolutionary constraint imposed by the mechanics of the eye's development; it is not without its costs, for in the very center of the primate eye natural selection has modified the retina by pushing the neurons to one side, creating a clearer path to the receptor cells.[11] These are but two examples of many kinds of constraints with which adaptation must compromise in optimizing fitness. One problem with defining fitness in terms of engineering design is that we rarely know the range of options that were available along the evolutionary path. Another is that the engineering problem may not have remained constant in time.

An alternative to defining fitness in terms of anatomy, physiology, and behavior is to measure fitness in terms of its contribution to future generations. Isn't that, after all, what fitness is supposed to be about? But here too we meet equally severe conceptual difficulties. How far into the future do we look to measure contributions? How do we separate reproductive success due to chance from reproductive success due to design? What, in fact, do we look for in future generations? Individual, sexually reproducing organisms do not reproduce themselves; no carbon copy exists in future generations. This is because their genes get shuffled and recombined with others during each reproductive event, but as a result no package of genes (genotype) reproduces itself either. With the exception of identical siblings, every sexually reproducing organism is genetically unique, and we cannot find duplicates of any organism or of its package of genes in succeeding generations.

One possibility is to assign values of fitness to entities that do occur in successive generations. Thus, population geneticists attach values of fitness to individual genes. But this exercise must average the fitness of the gene over all of the genetic backgrounds in which it occurs in the experiment, which is never all possible genetic backgrounds. Moreover, the fitness of a gene is also likely to be a function of the environment in which its bearer finds itself. This latter consideration is profoundly important and is one to which we shall come back in due course, but for now we need to deal more successfully with this notion of fitness.

For our purposes a more general if still somewhat intuitive and arbitrary measure of fitness builds on the idea that genes—albeit in interaction with the environment—specify a general design of the organism. That design—however it is defined in particular cases—is expressed (with variation) by individual organisms through their anatomy, their physiology, and their behavior. The success of the design (or more precisely, a variation of it in competition with other variants) can be assessed by measuring the contribution of progeny to successive generations. As chance events may compromise the reproduction of any individual, such a measure of fitness must be based on replicate measurements. Organisms are therefore fit in this evolutionary sense by virtue of appropriate design. How fit is the design? The fitness of a design is roughly in proportion to the frequency with which the design appears in successive generations when it competes with alternative variations. Fitness therefore depends on the genetic alternatives as well as the environmental conditions.

Design is one of those words that may suggest to the non-biologist a guiding hand and some measure of foresight or purpose. That, however, is not how evolution works, and the use of the word "design" should not be so construed. Nature does not anticipate. We speak of design as a convenient way to indicate that there is a reasonably effective match between the organism's features (its phenotype) and the requirements it faces in its normal environment. Gills are organs of respiration that enable fish to extract oxygen from the water in which they swim. Gills are well designed for the purpose in that they function efficiently, but gills are the product of natural selection and not the work of a draftsman and engineer. Mammals, on taking to the water permanently, did not have gills and have never evolved them. Whales and porpoises might have benefited if organisms were the result of conscious design, but they must instead come to the surface to breathe as do their land-bound kin.

Tallying offspring as a measure of fitness may be relatively simple to accomplish, but should we be concerned with abundance or with persistence of progeny? Posing that question calls attention again to the importance of environmental change. There is one situation where counting offspring to assess fitness involves a genetic simplification: If all offspring of any given female were genetically identical (forming a clone), then we could obtain information about relative fitness in a given environment by observing changes in the relative abundances of different clones. Clonal organisms are not uncommon. For example, aphids, weevils, lizards, and other organisms have parthenogenetic forms in which females give rise to exclusively female offspring, all of which are genetically identical to their mother. With such a mode of reproduction, clones that predictably increase their representation at the expense of others might well be deemed more fit. But clones that have a high rate of increase in one environment may have a high probability of extinction if the environment changes. If environmental changes occur on a time frame comparable to the generation time, abundance and persistence will be comparable measures. If, on the other hand, significant environmental changes occur over periods of time long compared with the generation time, then abundance at any particular time may not be a good measure of persistence. Natural selection operates without a crystal ball to tell it what environmental changes the future will bring, and organisms do not evolve in anticipation of future events. Fitness can only be specified in the context of a particular environment; phenotypes, and their underlying genotypes, that are currently most fit may lose their advantage should the environment change. But a corollary is that effective design is likely always a compromise, and lineages that persist may do so despite a selective disadvantage in the short term in particular environments.

Both survival and reproduction are components of fitness, but we shall defer further discussion of this matter until a later chapter, when we consider the great variety of life histories as evolutionary "designs."

Some recent contributions to evolutionary theory particularly relevant for the study of behavior

Inclusive fitness and kin selection. Fitness, in the Darwinian sense of reproductive success, implies that genes guide the creation of a phenotype that is designed to ensure their propagation into future generations. The genes of any individual occur not only in its offspring, however, but also in siblings and, with decreasing probability, the offspring of siblings, cousins, the offspring of cousins, and so forth. The concept of fitness therefore can be expanded to include descendants of near relatives, appropriately devalued as the relationship becomes more distant and the probability of having shared genes decreases.[12] This augmented count of shared genes in relatives is called *inclusive fitness*. To put some numbers on the probabilities: half of your genes are passed to your children; on average you share half of your genes with a full brother or sister; and a quarter of your genes will be present in your siblings' children or your own grandchildren. Thus, in general, natural selection can be expected to favor behaviors that enhance the reproductive success of near relatives. Interestingly, an individual animal's inclusive fitness may be larger, even at a cost to the reproductive success of itself, by acts that ensure the survival and reproduction of a sufficient number of close kin. Nepotism is therefore in the interests of inclusive fitness.

This issue was brought to the fore in part by the recognition that altruistic behavior—behavior that seems to benefit others at the expense of self—should fly in the face of natural selection. Natural selection should generate reproductively selfish behavior. Part of the solution to this seeming paradox is addressed by the concept of inclusive fitness, for if animals behave more benignly toward near relatives than they do toward strangers, their behavior is not really altruistic at all. They are indeed looking to their inclusive fitness. Evidence exists, moreover, that nepotistic behavior is quite common in the animal world. The social structure of bees and ants, in which thousands of sterile workers toil for the reproductive benefit of a single fertile queen, is based on a genetic system (males are haploid, possessing only one set of chromosomes) in which females share more genes with their sisters (3/4) than they would with their daughters (1/2). This is because every worker inherits *all* of her haploid father's genes but only half of her mother's. (This is now known to be an oversimplification, because although the queen bee has but a single nuptial flight, in that brief time she collects and stores the sperm of more than one drone. The average genetic relatedness of female offspring will therefore be somewhat less than 3/4. The genetic system of these insects nevertheless provides an explanation for their altruistic social behavior as long as the average relatedness of sisters is greater than 1/2.)

Although the development of the theory about inclusive fitness was driven by analysis of the genetic system of these social insects, the general concept is quite applicable to mammals. Some have questioned whether inclusive fitness can have any relevance to human behavior, for how does an animal, human or otherwise, know kith from kin? As it turns out, a variety of animals—bees, birds, mammals, including humans—have visual, auditory and olfactory mechanisms for recognizing close relatives, which they then relate to in a manner different from strangers. For example, Belding's ground squirrels are more tolerant of genetic relatives in adjacent territories than they are of total strangers.[13] The nepotistic interests of humans decrease with distance of relatedness and are conveniently calibrated by the Arab adage "Myself against my brother; my brother and myself against my cousin; myself, my brother, and my cousin against an outsider."

Finally, in a small group of related individuals, one might expect to observe natural selection for acts that benefit the entire group.

This phenomenon is known as *kin selection*, and it follows directly from the concept of inclusive fitness.

Just what is it that gets selected? We saw above that natural selection is a sifting of alternative genotypes, but the *process* by which this occurs involves the phenotypes. Genes are not selected on the basis of their physical or their chemical structure, but indirectly, by virtue of their phenotypic effects. Moreover, genes characteristically travel in battalions and regiments (genotypes) and participate jointly in the embryological development of individual organisms. Phenotypic effects of genes depend on which other genes are present in the organism as well as on a great number of environmental influences of varying specificity. The reproduction of organisms is the means by which genes replicate themselves into the future, and as Richard Dawkins[14] has put it, each time an individual organism starts to develop from a single cell, natural selection is presented with a new set of opportunities. The proximate means of natural selection is therefore largely a sorting of organisms.

This understanding clarifies a common misconception: Natural selection does not function "for the good of the species."[15] Successful replication of genes is the name of the game, in which organisms are the agents of the genes. Because organisms are the focus of selection, one cannot argue that selection fosters the interests of groups, except perhaps indirectly. Of course, if the group consists of closely related individuals, the genetic interests of the group and the individuals of which it is made up are largely coincident, but this is usually not the case.

The problem of group versus individual selection has been treated mathematically, and it is interesting to ask what conditions, if any, would permit the focus of selection to shift from individuals to groups. Group selection is a theoretical possibility if the rates of extinctions of groups are high relative to the generation times of individuals.[16] This might be found with small groups that engage in intense and lethal intraspecific competition, a situation that may well have characterized human populations from time to time.

The concept that microevolution results from the reproductive competition of individual organisms has powerful predictive value and will serve us well in the remainder of this book. There are a couple of additional caveats to be considered, however. As a general proposition, objects that are able to replicate or reproduce, that exhibit variation among members in their success at replicating, and for which some of the variation is heritable—in short, entities that possess the critical trilogy of Darwinian features—should evolve by natural selection. We saw earlier that the great unexplained excess of DNA in most eukaryotic organisms has led to the hypothetical possibility of DNA propagating itself within genomes with no phenotypic consequences. Should this prove to be a real phenomenon, then we shall have to recognize both that organisms are not the only level of organization at which selection occurs and that there are situations in which genes may compete directly.[17]

Second, evolutionary theory that simply *assumes* the presence of organisms has little to say about how organisms came into existence through the cooperative behavior of cells that all possess the same genetic endowment, and how cells were formed from a simpler soup of replicating nucleic acids. Life is a hierarchy of organizations building from molecules, cell organelles, cells, tissues, organisms, populations, species, and ecological systems of organisms. This hierarchical structure is the result of evolution, but evolutionary theory does not yet deal effectively with aspects of evolution that are not readily observed today. For example, it is likely that in the past the major action of selection has occurred at simpler levels of organization—molecules and cells—and that with the appearance of more complex entities in the hierarchy, lower entities fell under the control of higher, and the level of selection shifted.[18]

Finally, the concept of alternative levels of selection has been extended upward from organisms to species, but not without much disagreement. As described above, there is considerable reason to question whether under most conditions the microevolutionary forces that drive individual selection can be applied to groups. The fundamental issue posed by the idea of species selection seems to be whether the patterns of geographical expansion, retraction, and extinction of species that are seen in the fossil record are simply manifestations of the successes and failures of natural selection working at the level of individual organisms. The microevolutionists by and large assert that current evolutionary theory is quite capable of accounting for these paleontological observations, but the issue remains alive among a subset of evolutionary theorists.[19]

The question of which and how many levels of selection remains an important theoretical issue, and it illustrates how the scope of problems posed by evolutionary change is larger than the current theoretical construct has absorbed. Contrary to the impression left by many elementary texts, evolutionary biology was not put to bed by Charles Darwin; it remains a dynamic science with much to understand. For our purposes, however, it will be sufficient and largely correct to recognize that for the last several hundred million years most evolutionary change in the major groups of animals has been driven by reproductive competition between individual organisms.

Sexual selection. Every December on the island of Año Nuevo off the California coast, elephant seals begin to assemble for their annual cycle of reproduction. The bulls are great galoots, weighing several times as much as the females, and they spend the next two months on the beach in brief but violently aggressive encounters with one another in efforts to control access to the females. Only the most dominant males are successful. The result is that although most of the females older than three years become pregnant, there is enormous variation in the reproductive accomplishment of males. From field records of successful copulations, it appears that in one season four percent of the males were fathers to eighty-five percent of the next year's offspring.[20]

Elephant seals present a dramatic example of a phenomenon noted by Charles Darwin. In many species the two sexes differ greatly in morphology and behavior. Darwin further observed that individuals of one sex, usually males, compete with one another for mates, and that individuals of the other sex exhibit some discrimination in choosing a mating partner from among the competitors. He recognized that this competition for mates could result in evolutionary change and termed the process *sexual selection.* The idea of sexual selection is therefore not a new addition to the corpus of evolutionary theory, but during the last twenty years the concept has been fortified with additional insight.

The theory of parental investment. The great diversity in sexual dimorphism, mating practices, and life histories of sexually reproducing organisms is one of the most obvious and interesting results of the evolutionary process. The theory of *parental investment*, due largely to Robert Trivers,[21] is an effort to formulate the major forces driving the underlying sexual selection. The central idea starts with the premise that there is some cost to an individual organism (in terms of time, energy, increased chance of earlier death) in producing offspring. Parental investment is defined by

Trivers as "any investment by the parent in an individual offspring that increases the offspring's chance of surviving (and hence reproductive success) at the cost of the parent's ability to invest in other offspring." To anticipate a more detailed discussion in the next chapter, one form of parental investment is in the production of germ cells; many sperm can be produced for the same metabolic cost as a single egg. By this measure, the parental investment of females is greater than that of males. The differential cost is greatly magnified in species (such as mammals) where the fertilized egg must be nurtured by the female for an extended period of time before the birth of the offspring. On the other hand, the parental investment of males may be greatly increased by defense of breeding territory, care of young and other activities. In fact, in some species the parental investment of males associated with care of the young may exceed the total parental investment of females, with the result that the usual patterns of competition and choice associated with males and females are reversed. We shall see some specific examples in the next chapter.

Implicit in this definition is the supposition that the costs of parental investment are of sufficient magnitude to constrain total reproductive output. If a bird lays too many eggs, the ability to feed the extra young may jeopardize the chances of rearing any of the offspring to maturity. Depending on the life history of the species, the added risks of reproduction (e.g., susceptibility to predators, death at childbirth) can compromise the lives of other offspring.

When there is a great disparity in the parental investment made by the two sexes, the sex that makes the greater parental investment tends to become a limiting resource in reproduction. Several things then follow. First, there is a competition among individuals of the opposite sex for breeding. This competition frequently leads to selection for features such as larger size and more aggressive behavior. In other words, differential parental investment lies behind sexual selection. Second, if individuals of the sex making the greater parental investment exercise some choice of mating partner, there can be selection for greater parental investment by individuals of the opposite sex.

The resulting evolutionary results are diverse. Moreover, because there are a number of potentially conflicting selection pressures at work, specific outcomes are hard to predict from first principles. For example, in a species where males compete for access to females but female choice influences the outcome, females might be expected to chose males that would make the greater parental investment (e.g., be least likely to desert while the young need care). There would then be selection among males for both an increase in parental investment and a means of advertising it to females. On the other hand, selection for an increase in parental investment would likely be opposed by selection for male behavior that maximized the number of females mated. Furthermore, the resulting outcome—the mating system—would likely be constrained by other aspects of the life cycle. For example, the nature and extent of male parental investment on which selection could operate would depend on whether the species has slowly developing offspring. We shall expand on these matters in the next chapter.

Evolutionarily stable strategies (ESS) embrace a mode of thinking about alternative phenotypes that is borrowed from mathematical game theory. The word "strategy," like "design," which we met earlier, does not imply any conscious planning. An ESS is a phenotype (in practice most often a behavior) that, if adopted by all members of a population, cannot be displaced through the action of natural selection by any alternative mutant strategy.[22] Suppose two elephant seals have an aggressive encounter over a mate. Each has the option of pressing the case or retreating from the field. Reward follows from the former, safety from the latter. But the former also entails a cost, which might include serious injury. The evolutionarily stable strategy is one that can weigh the probabilities of cost and payoff, the reproductive consequences of each, and escalate the aggression only when the payoff (in potential fitness) exceeds the corresponding costs. None of this language should be taken to imply that animals are using an abacus to calculate anything, but the conceptual framework comports qualitatively with many features of animal behavior. It provides a reason, in terms of individual selection, why aggressive encounters between bull elephant seals or horned animals usually terminate short of serious injury, or why the clutch size in birds is adjusted to the resources required to feed the young and not the egg-laying capacity of the female.[23] Evolutionarily stable strategies are not easy to quantify, partly because behavior frequently presents so many options, and partly because of the assumptions that must be made in assigning numbers to costs, benefits, and probabilities. It remains, however, a useful way to think about evolutionary design for fitness.

4

Reasoning about Ultimate Causes of Behavior

> But as men are most capable of distinguishing merit in women,
> so the ladies often form the truest judgements upon us. The
> two sexes seem placed as spies upon each other, and are fur-
> nished with different abilities, adapted for mutual inspection.
>
> *The Vicar of Wakefield*
> Oliver Goldsmith (1776)

> We cannot fight for love, as men may do; We should be
> wooed, and were not made to woo.
>
> *A Midsummer Night's Dream* (II,i)
> William Shakespeare

In the last chapter we saw that during the previous twenty-five years our understanding of biological evolution has been deepened by several new or newly clarified concepts. Specifically, the following four stand out in importance for the evolution of behavior. First is the realization that in the vast majority of cases natural selection takes place at the level of *individuals* and not groups.[1] Thus, arguments that such-and-such structure, process, or behavior occurs "for the good of the species" are generally incorrect. This is a point that was clearly understood by Charles Darwin but for many biologists was only brought into sharp focus by George Williams's *Adaptation and Natural Selection*, published in 1966. Second is the concept of *inclusive fitness*, which recognizes that copies of many of an animal's genes reside in close relatives and that in certain circumstances fitness can be enhanced by seemingly altruistic acts extended to others.[2] Third is the theory of *parental investment*, which has far-reaching implications for mating behavior and the relationship between parents and young.[3] And fourth is the concept of *evolutionarily stable strategies*, which arises from an application of the theory of games to the evolution of behavior.[4]

We are now going to examine in more detail some arguments that are based on these ideas. We shall use for our prime example a matter that has very broad and general importance in the animal kingdom, that is central to reproduction and fitness, and yet whose application to the human condition is frequently misunderstood or distorted: the relationship between the sexes.

What is the meaning of sex?

First, however, let's consider a fundamental question that we alluded to earlier: Why does sex exist at all? When we speak of sex in ordinary speech we usually mean male or female gender or the physical act of mating. Sex also refers to something more basic: the mixing of parental genes to produce a new genotype.

The very existence of sex presents a biological conundrum. What makes it a conundrum, however, starts from the conclusion, independently reached, that individual organisms and not larger groups or species are the principal focus of natural selection. The paradox is as follows: If the fitness of the parent has been well tuned by the history of selection of the parent's ancestors, the shuffling of genes (by independent assortment and recombination) that takes place in the formation of eggs and sperm together with the new diploid combinations created when the egg is fertilized should break up successful assemblages of genes and generate genotypes with substantially reduced fitness. Producing poorly fit genotypes is not only a waste of reproductive effort, it is also a loss that would not be incurred if the parent were to reproduce asexually. Furthermore, in species with negligible parental investment by the male, females would maximize their reproductive output if they relied upon asexual reproduction and produced only daughters.

What, therefore, is to be gained by sex? Sexual reproduction might provide a way of probing for even better combinations of genes. Which mode of reproduction is then best? The answer seems to depend on the circumstances, but the problem is not thoroughly understood. Ordinarily, if natural selection has been effective in a lineage, better combinations of genes for the existing environment are relatively improbable. But if a deleterious mutation should arise in the parent, sexual reproduction provides an array of genotypes among the offspring that natural selection can then edit. In contrast, when a mutation arises in an asexually reproducing line, unless it is removed by another mutation it will be found in all offspring and in their descendants. If individuals bearing the mutant gene encounter strong negative selection, the gene can drag into oblivion all of those individuals, along with the rest of their other genes.

With asexual reproduction, selection will occur between clones, and under certain conditions the rate of evolution will be substantially slower than is possible in the presence of recombination. Sexual reproduction can lead to faster evolutionary change if the breeding population is large enough to bring together favorable new combinations of genes at rates that exceed the successive accumulation of favorable mutations in asexual lineages.

Sexual reproduction will be favored where there is strong competition between individuals. This condition may be met if the organisms are at high density and have limited capacity to disperse. Similarly, sexual reproduction may increase the number of offspring able to occupy variant niches in the habitat, also a more likely necessity at high population densities.

A further potential advantage to producing a variety of genotypes arises when the environmental future is variable and uncertain. The organism does not, in fact cannot, plan for future generations, and like a poker player whose luck is not limited to the quality of the first five cards he is dealt, sexual reproduction gives the organism at least a chance of putting forward an assemblage of genes better suited to the selective competition that will be encountered. But for sexual reproduction to prevail, this benefit must exceed the costs of sexual reproduction that were enumerated above. The "Red Queen" hypothesis (named for the character in Lewis Carroll's *Through the Looking-Glass* who explains to Alice that in her world " . . . it takes all the running *you* can do, to keep in the same place") is a variant of this idea whereby it is suggested that the ever-changing nature of environments puts a premium on the genetic diversity of offspring just to keep the lineage going.[5]

Fluctuating environments will assert a maximum premium on sexual reproduction when environmental changes actually *oppose* the current adaptational state and do so relatively frequently. This condition is most readily imposed not by the physical environment, where fluctuations are more likely to be random, but by the biotic environment.[6] Predator and prey, host and parasite are in states of adaptive warfare in which an improvement in one member issues a counterchallenge to the other. The occurrence of asexually reproducing organisms tends to support the paramount role of other organisms in generating the environmental instability responsible for sexual selection. Sexual reproduction is more common in parasites than in closely related but free-living forms. Asexual reproduction is more common in fluctuating, disturbed, and ephemeral habitats, at high latitudes relative to the tropics, and in fresh water rather than marine environments. These general correlations at first seems surprising, but they suggest that the physical environment is not in fact the primary causal agent in determining whether reproduction is sexual or asexual. Asexual reproduction correlates better with conditions in which there is a premium on a high rate of reproduction and where there are minimal interactions between individuals or between species.

The fact that some organisms reproduce sexually, others asexually, while still others switch modes indicates that there are advantages and disadvantages associated with both systems. The existence of sexually and asexually reproducing organisms is but one of numerous examples we shall encounter of natural selection's traveling multiple paths. The origin of sex is also a realm of evolutionary theory in which there is considerable intellectual uncertainty and ferment.[7]

The fundamental significance of parental investment

In parts of what now follows I have quite consciously paraphrased Richard Dawkins,[8] an Oxford biologist whose vivid and inspired writing on evolution has successfully reached a general audience. I have done this, however, not in criticism of his lucid style, but to illustrate a point about the use of language in discussions of ultimate cause. Many readers may find portions of this description troublesome, because a number of the common English words and phrases will seem to be invested with too much meaning. I have flagged some of the more likely examples with quotation marks, and I shall return later in the argument to a further discussion of the language.

Sex can exist in bacteria and fungi in the form of different mating types, yet without the morphological differences we recognize as male and female. How do we decide which sex is female and which is male? We recognize that in general the two sexes have adopted different "strategies" for the manufacture and dissemination of their germ cells. One sex makes relatively fewer cells (eggs) and invests more of itself in each. This is what Dawkins has termed the "honest" strategy, and the individuals who practice it we call females. The other sex makes more but smaller and more mobile germ cells (sperm) whose task it is to find the eggs. Unlike the egg, each sperm has little to contribute to the zygote but its genetic material, its DNA. This is the "sneaky" strategy of Dawkins, and its practitioners we call males. To quote Dawkins again: "Female exploitation begins here." Let us explore what this means.

For many species, parental investment only starts with the formation of germ cells, and there is a great deal of species difference in the disparity of investment made by the two sexes. In fish (where fertilization is external), both sexes cast their seed into the water and the differential investment in eggs and sperm is probably not large. In some species the young receive no parental care at all; the eggs are deposited in the water, fertilized by an attending male, and the fry fend for themselves on hatching. In other cases, occurring in ponds or coral reefs (where the habitat is structured and demarcated), males defend territories and may frequently protect their young, keeping them together, sometimes by sequestering them in the males' mouths.[9]

Among birds and mammals the situation is different. Because fertilization is internal, the female is stuck with the zygote in a way the male is not. Female birds put a lot into their eggs, not just figuratively but in fact. Although the eggs of female mammals are much smaller than those of birds, the embryo is nurtured internally and the young are fed after birth by milk secreted by the mother. All of these features place an energetic burden on the female that the male shares only indirectly, if at all. When the female's parental investment is much more extensive than the male's, the female is potentially capable of contributing genes to fewer offspring than the male is. Under these conditions, females are, for males, a "resource," and selection tends to favor patterns of reproduction in which males attempt to inseminate many females. For females, on the other hand, there is usually not an equivalent advantage to be gained by multiple matings. Moreover, because for the males the females are a valuable resource, they frequently become the object of competition between males. This competition expresses itself in various ways, often in aggressive interactions between males, by trends toward sexual dimorphism—with the males becoming larger and more aggressive than females, and a tendency for males to try to coerce females into mating and to prevent them from mating with other males. As mentioned previously, the ultimate causes of sexual dimorphism were recognized by Darwin as *sexual selection*. Sexual selection is a result of differential parental investment.

Successful genes are those that are propagated into subsequent generations. Because the genes of males and females are packaged in different bodies, they must use somewhat different strategies to achieve this end, and there is therefore some conflict inherent in the interests of the two sexes. Each parent makes essentially the same contribution of genetic material to the zygote, and there would be some advantage to each if the other could finish the job of rearing the young, freeing the first to start another round of reproduction. But as the total initial investment of the female (particularly a bird or mammal) is greater, we would expect to find males much more likely than females to desert once fertilization has occurred, leaving the female to complete the task. On the other hand, desertion is an evolutionarily acceptable strategy for a male only if there is a reasonable likelihood that the female will be successful in rearing the young without his help. Moreover, the female's genetic interests will be served by processes that decrease the chances of desertion and increase the parental investment of the male.

If the male's reproductive interests are fostered by multiple inseminations, a tactic for the female is to defer her choice of partner until after a courtship period in which males compete in demonstrating their capacity for parenthood. In short, the females become "coy." For example, in many species of birds the males stake out territory sufficient to support the young, start building nests, and announce their presence with songs and bright feathers before mating. This process also serves the female by increasing the male's parental investment, a tactic that is effective if it involves him until it is too late in the breeding season to seek another mate. Moreover, it allows the female to choose a mate "that is in the best interests of her genes."

Because the reproductive interests of males and females are never identical, there are inevitable conflicts. Even when the parental investment of the male is large, for example in guarding territory or in feeding young, there is still some potential gain to be had by "philandering," even short of deserting. It is in the interests of the female, however, to minimize the time and energy her mate expends on creating offspring who do not share her genes, and as we have described, the female can look to her interests by seeing that the male spends time and energy wooing and by being selective in her choice of mate. Seen from the perspective of the male's genes, though, if he is to make a substantial parental investment, it pays him to maximize the odds that he is the father of any offspring. Whenever fertilization is internal, as in birds and mammals, the male can never be certain of paternity unless he pays great attention to his mate's activities. Consequently, it becomes particularly important for males with a large parental investment to guard their mates to prevent cuckoldry. Cuckoldry can be in the interests of the female, however, if a "better" male should be transiently available.

There is another reason why females might be expected to be more "fussy" than males in selecting mates. As they have more invested, they have more to lose should the offspring be genetically inferior. Assuming incestuous relationships are initiated by the older partner, mother-son incest should therefore be rarer than father-daughter incest,[10] a prediction that seems to be fulfilled in human populations.

Another pattern of courtship involves displays by males, such as the bowers of bower birds or the struts and accessory plumages of grouse and peacocks. These male behaviors, which appear to our eye irrelevant to the practical business of rearing young, seem to convey to the females some general information about the vigor and general genetic worth of potential mates. It is in the interests of the female's genes if she can produce sons who themselves are successful in siring offspring. "Sex appeal" therefore becomes a desirable end in itself.

As females are generally the reproductive resource, competition for mates should be more frequent between males than between females. Competition between females is indeed less obvious, and it tends to be of a different sort. In a mating system where a few males are inseminating most of the females and there is little further paternal investment, not only is there no reason for females to compete for mates but it is actually in the reproductive interests of every female to mate with the most fit males. Under these conditions, therefore, polygyny (where one male mates with several females; see below) can actually work to the genetic advantage of the females. Food and a safe habitat present a different kind of challenge to the females, however, and active competition for and defense of these resources by females is common in many species. In some primates there is a dominance hierarchy among females, and the offspring of dominant mothers enjoy the results of the mother's social status, for example in access to food.[11,12]

In those animal societies (e.g., langurs, lions) where exclusive breeding opportunities are a temporary privilege of any male, usurpation may be followed by the new male's trying to kill the infants that were sired by his predecessor.[13] This behavior has struck some observers as an aberration on the mistaken premise that it is maladaptive, but this interpretation assumes that adaptive behavior must be for the good of the species. In these instances nothing could be more in the interests of the new male and his genes than to get the females ready for him to inseminate as quickly as possible.

An argument about evolutionarily stable strategies

Dawkins has presented a nonmathematical description of the consequences of various reproductive behaviors for an evolutionarily stable strategy.[14] In brief, if all females were "coy" and required a demonstration that males were "faithful" before mating with them, the system would be unstable because it would be open to females who cheated by being "fast" and not enforcing a period of courtship. The presence of "fast" females would open the way for "philandering" males, who would otherwise be excluded. The presence of "philandering" males would in turn drive up the frequency of "coy" at the expense of "fast" females. What results from this kind of analysis is a mix of behaviors in the population (and over time) that is stable in the sense that it is immune to perturbation by changes in the frequency of any of the behavioral forms. What the actual proportion of "coy," "fast," "faithful," and "philanderer" are depends on the assumptions one makes about the payoffs and penalties to reproductive success that are associated with each behavior. It is important to realize, however, that this kind of modeling does not assume that each behavior is rigidly fixed; an individual female could be "coy" part of the time and "fast" the remainder, and the conditions of the model would be fulfilled. We shall see subsequently that behavioral plasticity of an equivalent degree is readily observed in nature.

Concerning the language

The preceding several pages convey some of the flavor of evolutionary arguments about the ultimate causes of behavior. The reader who is unaccustomed to thinking along these lines may well be disconcerted by some of the language I have used: "successful" genes, evolutionary "strategies," an animal's "confidence" in paternity, "philandering" male animals, and creatures who "choose mates that are in the best interest of their genes." These kinds of descriptions have been criticized because they suggest moral judgments by animals and conscious foresight by molecules. The criticism is understandable, because language is supposed to be able to convey meaning unambiguously. Biologists, for their part, bring their own background and understanding to the problem, and frequently find such interpretations so silly as not to require comment. The result is often a serious failure of communication.

We must establish a common understanding of this matter before we proceed further. Nothing in these arguments about the adaptive significance of various patterns of reproduction is intended to address the *proximate* mechanisms that determine mating behavior. For example, calling behavior "coy" is simply a shorthand way of saying that the female does not copulate until time has passed and more information is available to her. It implies nothing about the cognitive processes by which that information is assimilated, her emotional state, or any other property of her nervous system. Such things are important, but they are not directly relevant to statements about the overall effects of "coy" as opposed to "fast" behavior on the transmission of the female's genes to the next generation. The argument obviously assumes that there are some mechanisms by which "coy" or "fast" behavior can be expressed, but the details of those proximate causes are generally not of central concern when one is thinking in terms of ultimate cause and evolution. Similarly, to say that the male has "confidence in paternity" simply means that the animal is involved in a mating system in which the probability is low that his mate will be inseminated by another. Whether he frets about the problem is another matter. Clearly males of at least one species do.

The alternatives to a breezy style of description of animal behavior can be cumbersome or involve the creation of a special jargon. To date, evolutionary biologists have had relatively little trouble communicating with one another without a great deal of the latter, but they sometimes have a serious problem in reaching beyond their circle. Bridging this communications gap will require more sophistication in distinguishing between proximate and ultimate cause than our educational system now provides most students—or is likely to provide in the near future. Greater efforts on the part of biologists will be needed to convey the power, importance, and beauty of evolution.

Finally, although evolutionary biologists are often not particularly concerned with proximate cause, social scientists and behavioral physiologists clearly are. Much of the remainder of this book has to do with the manner in which explanations of proximate and ultimate cause can illuminate each other. But for the moment this is getting too far ahead in the story.

Mating systems

The concept of differential parental investment holds the key to understanding various mating systems: monogamous, polygynous, and more rarely, polyandrous. What has been described to this point as the basis for sexual selection is a polygynous system, in which one male mates with a number of females. Characteristically, in polygynous systems of birds and mammals, most of the females of reproductive age become inseminated and attempt to reproduce. Their reproductive success is determined more by the resources available to them—a suitable breeding site relatively safe from predators and with access to sufficient food to feed the young—than it is by the availability of males. The reproductive success of males, on the other hand, is more likely to be limited by competitive encounters with other males or with other dangers, for example, predators, to which the male is exposed during conspicuous and persistent courtship behavior. As a result, the variability (again, the variance) of the reproductive success of males tends to be greater than that of females. As with elephant seals, one or a few males may achieve most of the matings in a strongly polygynous system in which there has been vigorous sexual selection, and some males may be cut out of the picture entirely.

The Canada goose makes long-term pair-bonds. In summer the geese travel to the far north to breed, returning south in the winter to find open water and a supply of food. Their honking vee formation is a nostalgic reminder of the changing seasons, and at least in our culture, belief in their monogamous mating system seems to stir human emotions with a similar power. In fact, most birds are monogamous, at least in the limited sense of forming pairs within a single breeding season, whereas most mammals are to some extent polygynous. Only about four percent or fewer of mammalian species approach monogamy. What circumstances tilt the mating system toward monogamy? Simply put, the less disparity there is in parental investment, the more likely the mating system will tend toward monogamy. It has been argued that the difference between the monogamy of most birds and the polygyny of most mammals follows from the fact that female birds deposit their developing embryos outside of their bodies in large, shell-covered, yolk-containing eggs, whereas female mammals gestate the embryos internally, nourish them before birth through a special structure called the placenta, and following birth feed them milk, also produced by the mother.[15] The mammalian form of maternal investment is enormous, and for many species the paternal role is correspondingly reduced. Following insemination of the female, to the extent that young will develop successfully to weaning without any further paternal input, the reproductive interests of the male are best fulfilled by seeking other females to inseminate. For birds, on the other hand, particularly those whose young hatch from the egg in a relatively naked and helpless state, sustained parental investment by the male may be an absolute necessity in order just to feed the gaping mouths. As long as the bonds remain stable enough to rear young successfully, strict monogamy may nevertheless be compromised by sneak fertilizations. These are always to the advantage of males, and can be equally advantageous for females if a genetically superior male should present himself. Strict monogamy, defined in terms of sexual encounters, is probably more a human ideal than it is a common biological reality.

The nature and degree of polygyny vary greatly in their expressions, even among related species.[16] Frequently they appear to be tuned to other aspects of life history. Among the hoofed mammals that graze and browse for sustenance, there are many different examples. The dik-dik is a diminutive African antelope that lives in brushy country and hides from predators. Dik-diks are found in seemingly monogamous male-female pairs. At the other extreme are the largest grazers like the eland and buffalo that congregate in herds where individual males fight to control the copulations of as many females as possible. In between is the impala: modest in size but considerably larger than the dik-dik. Impalas move in herds and flee from predators. When breeding, adult males establish territories and wait for a group of females and immature males to pass through. The male's life then becomes a frantic rush to copulate with as many females as possible while chasing young males. The male has to scramble for his reproduction rather than having to fight for it, and anyone who witnesses this performance is likely to feel pity for his plight as well as wonder how he accomplishes anything.

The basic social unit of the hamadryas baboon of Ethiopia is a single male and a harem of several females. These units may congregate in larger assemblages, forming bands and troops. The males jealously guard their females, biting them on the neck if they stray, and keeping other males from mingling with them. Other species of baboons from farther south in Africa have a somewhat different social structure consisting of troops of two or three dozen individuals of both sexes, adults and juveniles. Females remain in the troops of their birth, whereas there is some exchange of males between troops. The males have a dominance hierarchy that is reasonably stable over time but subject to rearrangement as individuals come, go, mature, and age. The frequency of copulations that males achieve corresponds with their social rank; higher ranking males are more successful in mating females. The system is not as exclusive as in the closely related hamadryas baboon, because a given female is not monopolized by a single male.

The primates as a group show great variation in social and mating systems. The orangutan (an ape) and a number of nocturnal prosimians (tree shrews, lemurs, lorises, galagos, and relatives) are solitary. Both the Old World and New World families of monkeys contain examples of species that congregate in single-male, multi-female bands or multi-male, multi-female bands, along the lines described for baboons.

Our closest primate relatives, the chimpanzees, usually live in multi-male, multi-female groups, but some all-male groups also exist. When individuals transfer between groups, they are usually females, an unusual pattern for mammals but one also shared with gorillas and a few species of monkeys. The human social structure based on male kinship employing exchanges of women may therefore have its antecedents in the social organizations of the great apes. At one time it was thought that chimpanzees were indiscriminate philanderers, and it was hypothesized that male-male competition occurred primarily at the level of sperm production.[17] (A chimpanzee produces more than twice as many sperm per ejaculation than a man, and twelve times more than a gorilla.) More recent field observations, however, suggest that copulatory successes of male chimpanzees are not uniform.[18] Frequently at the time of estrus a female effectively pairs off for a few days with a single male. Higher ranking males are more successful in monopolizing females in this manner.

The theory of parental investment is supported by cases in which the roles of males and females are to some extent reversed.[19] In a family of fish that includes the little sea horse, the female transfers her eggs to a pouch on the male, who then gestates and nourishes the developing young. Sexual selection has been at work on these fish, for the females are not only the more brightly colored but also the active wooers in courtship.

Phalaropes, a small group of shorebirds, show a similar reversal in the usual pattern of sexual selection. The females are larger, brighter, arrive on the breeding grounds first, and guard the males from other females. The males, for their part, take sole responsibility for brooding the eggs and newly hatched chicks. Why have male phalaropes become the limiting reproductive resource, with the greater parental investment? A plausible explanation for this phenomenon draws on the breeding ecology of shorebirds. These creatures nest on the ground in open country, usually in the far north. The breeding season is short, and the nests are subject to predation. Since the young hatch able to feed themselves (like ducks and geese and unlike familiar songbirds), both parents are not required once the eggs are laid. If the female can defect and find another mate, she does so. In fact polyandrous mating (one female, several males) is practiced by other species of shorebirds that experience similar environmental problems and share the same precocial pattern of development. Polyandry is relatively uncommon among vertebrates but it is an exception that supports the rule of differential parental investment. We shall describe in a later chapter the special and equivalently rare case of polyandry in human societies.

Life-history strategies

The mating systems monogamy, polygyny, and polyandry represent but one measure of reproductive diversity. Some animals (and plants) concentrate their reproductive energies in a single but frequently massive effort. The butterfly characteristically lays her eggs and dies; the migratory salmon returns to the site of its birth only to terminate its life after an arduous upstream swim as part of a crescendo of reproductive exertion. Other species, however, reproduce repeatedly. The queen honeybee does little but lay eggs for several years, and birds and mammals generally reproduce seasonally, often more, sometimes less frequently. In animals, the capacity for repeated reproduction appears to be a prerequisite for the evolution of social organizations.

Life histories can also be characterized on a related dimension. At one extreme are species that produce many offspring, develop rapidly, exploit habitats quickly, and frequently have means for dispersal to distant sites. They frequently are found in unstable environments such as puddles or forest clearings, and there is a premium on at least some of the progeny finding suitable new places to live and reproduce. At the other end of the spectrum are longer-lived species that occupy more stable habitats, generally have larger bodies, develop more slowly, and reproduce repeatedly but have relatively few offspring. The former depend for reproductive success on a large number of offspring and often on the ability to colonize new areas. The latter depend more on their ability to exploit the habitat in which they live. Viewed another way, they also represent two extremes of parental investment, the first in fecundity, the second in nurture.

The two extremes were originally called, respectively, r-selected and K-selected species, a lifeless terminology that refers to parameters in the equations for growth of populations. The original concept was that some species have evolved with less emphasis on fecundity and more on competitive efficiency in the extraction of energy from their environment. As a consequence, in these latter species population sizes would tend to be more stable over time, but the densities might approach more nearly the carrying capacity of the environment. At the risk of oversimplification, this theoretical basis for r- and K-selection has been criticized on the grounds that r (the intrinsic rate of increase of the population) might reasonably be the object of selection, but K (the carrying capacity of the environment) is not a parameter of the same quality.[20] Regardless of how one formulates causal evolutionary hypotheses, however, animals do display this variety of life histories, and it is useful to think of this diversity as different designs for fitness, or more specifically, as different strategies of parental investment. We see once again that there are multiple solutions to the evolutionary challenge of reproduction.

Long life, slow development, and extensive parental care all require attention to garnering food resources and a greater need for the individual to ensure its own safety over extended periods of time. So we might expect to find in such species evolutionary changes that have increased the effectiveness of sensory detection and efficiency of exploitation of energy resources, including behavioral corollaries such as territoriality and sociality. With this form of life history, prolonged survival becomes necessary for successful reproduction, and with fewer offspring and a long period of development, mechanisms to forestall premature demise are also necessary. Behavior comes to exercise greater control over the rates of both births and deaths. In later chapters we shall elaborate on the idea that much of what is referred to as human nature is understandable as a consequence of the evolution of a long-lived, slowly developing, resource-requiring, mildly polygynous social primate that also happens to be highly intelligent.

What about the mating behavior of humans?

The origins of the relationships between the sexes in human societies have received much attention in recent years. Evolutionary hypotheses about male domination have been advanced, based on observations of contemporary hunter-gatherer cultures and on supposed divisions of responsibility in hunter-gatherer societies during the early evolution of humankind. To some extent these hypotheses recognize biological realities such as lactation, but they nevertheless remain narrowly based. As there is very little direct evidence as to how such societies were organized tens of thousands of years ago, it becomes possible to assert that they were truly egalitarian and that the domination of women is a cultural phenomenon that came later.[21] This is both fashionable and facile, and as others have sensibly argued, the very common asymmetry between the sexes of animals means that arguments about the causes of sexual selection should not be based on the characteristics of any one species.

The results of sexual selection in humans are evident to those who wish to see. The mild sexual dimorphism of body size and strength is accompanied by competition between young males for women, a greater variance in reproductive success among men than women, extensive efforts of men to control the reproductive destinies of women, a greater tendency of men to seek multiple partners, rape as a coercion of women by men and not vice versa, and prostitution as a female profession. Human males are capable of nurturing their offspring to an extent unparalleled among mammals, but that has not neutralized the force of sexual selection. Women still make the greater parental investment, and the evolved reproductive strategies of men and women are not congruent.

Among the primates, about eighteen percent of the species are seemingly monogamous. Where do humans fit in this picture? Some regularized pairing of men and women akin to marriage is so widely distributed across cultures as to be a characteristic of the human species. Features of *Homo sapiens* that contribute to this practice include slow development of the young with correspondingly great parental investment, long lifetimes, and elaborate social structures that are made possible by the cognitive capacity of the human brain. But what do we see in human mating practices that fits into the larger comparative and evolutionary picture that we have been painting?

By one estimate, about eighty percent of human societies are at least mildly polygynous and twenty percent monogamous.[22] These numbers alone, however, are misleading. Monogamous unions are in fact the most prevalent, even in societies where polygyny is acceptable. Polygyny is made practical when one man is able to accumulate sufficient resources to support more than one wife. Among the !Kung San people of Africa, whose traditional way of life is probably the clearest reflection still available of a preagricultural, hunter-gatherer culture, about ninety-five percent of the pairings are monogamous.[23] And in Western Europe and North America, where monogamy is decreed by law, divorce and remarriage effectively relax the formal imposition of monogamy.

The idea that polygyny is associated with accumulations of wealth and power is documented by an interesting analysis performed by Laura Betzig.[24] She compiled data on 104 autonomous societies from different places in the world and from different times in history with an eye to comparing information on polygyny and the means by which conflicts were resolved. She focused on a subset of twelve societies that exhibited the most complex political structure, with four hierarchical levels of organization. All were characterized by a despotic political organization, with vast power vested in a chief or king who settles disputes, frequently to his personal advantage. In all twelve, punishment of the aristocracy was less harsh than that of commoners. Interestingly, the concentration of power and despotic behavior correlates with extensive polygynous control of women, sometimes in such vast numbers as to preclude the potentate's sexual activity with all of them. In Incan society, access to wives correlated throughout the political structure. Depending on where a man stood in the formal hierarchy of principal persons, governors, administrators, and petty chiefs the system might provide "for their service and multiplying people in the kingdom" women in the numbers of 50, 30, 20, 15, 12, 8, 7, 5 or 3. Clearly a substantial number of men at the bottom of the hierarchy must have been without wives. This is a condition that ordinarily can be expected to generate considerable unrest and requires a despotic political structure to keep it from erupting.

Does polygyny increase male fitness? It certainly does for those males who hold resources and power. In all of the societies in Betzig's study, polygyny is associated with "a high degree of differential reproduction." This indicates why, in evolutionary terms, resources and power are important ends in themselves, a matter to which we shall return in later chapters. The promiscuous tendencies of men would seem to be obvious, and in Western cultures they have also been documented by attitudinal surveys.[25] Eight or nine times as many married men as women—in total, nearly half of the men—would like to engage in extramarital sex. The same individuals also held the view that men have a greater sexual desire than women.

The sexual interest of male mammals can frequently be aroused by novel females. Are human males an exception? This phenomenon is called the Coolidge effect,[26] supposedly in honor of a Presidential visit to a poultry farm. According to the story, Mrs. Coolidge, out of earshot of her husband, inquired whether a rooster could copulate more than once a day. On being told "dozens of times a day" she requested that this information be conveyed to the President. When this exchange was reported to Mr. Coolidge, he asked, "Same hen?" On hearing "no, a different one each time" he is alleged to have said, "Tell that to Mrs. Coolidge."

Monogamy is a mating practice that increases the parental investment of the male, which is to the advantage of the female. Although any offspring will have the female's genes, we have seen that the male cannot have the same certainty as the female about the presence of his own. The female's genetic interests are fulfilled if the male sustains his parental investment in her children, but the male's interests are furthered only if he is their father. The male may therefore gain in fitness by casual liaisons with other females, but he loses heavily (in terms of Darwinian fitness) if he invests in another man's offspring at the expense of what might have been his own. On the other hand, female infidelity has a genetic cost only if she loses the support of her mate, and under some circumstances infidelity could have a genetic benefit. Monogamy increases the male's confidence in paternity, but it does not make it certain. It is therefore no accident that through history human males have expended much effort in trying to control the reproductive activities of females.

There are numerous manifestations of this asymmetry, both cultural and emotional.[27] Female chastity is a virtual commodity in many societies. Unmarried women are chaperoned, veiled, hobbled, or otherwise protected or mutilated so that their value as future wives will not be compromised, and a transgression on their part can irrevocably soil the family honor. Violence induced by male sexual jealousy is a familiar theme in literature and life, and infidelity is the principal cause for the killing of wives by husbands. Until very recent times, most cultures have had a double standard regarding infidelity in that the crime is defined in terms of the marital status of the woman. The cuckolded husband is the aggrieved individual and is frequently excused of violence if he takes the law into his own hands. Margaret Mead's myths to the contrary notwithstanding, male sexual jealousy is a significant source of violence everywhere in the world.[28]

Martin Daly and Margo Wilson,[29] in a fascinating analysis of homicide statistics, have provided a great deal of insight into matters usually left to sociologists. Contemporary urban violence in the United States frequently seems to us now to be the result of drug wars, which to a large extent are also conflicts over resources. They are also largely conflicts between young men with otherwise limited economic prospects. In another generation, as indeed seems to have been the case for centuries in Europe and America (for which historical studies exist), a substantial fraction of all homicides involved young men fighting over utter trivia like a spilled glass of beer, a minor gambling debt, or a perceived insult to pride or honor. This is but one manifestation of what appears to be a need to demonstrate one's manhood, and it seems particularly prone to flare into violence where the participants have strained economic circumstances with little likelihood of improvement. But why should men and not women behave in this statistically predictable and dangerous way? What is it that drives the psyche of the young male to take such seemingly foolish risks? As the phenomenon is widespread and recurrent and the stakes frequently so high, the explanation of ultimate cause must involve something quite important to males. The most likely answer is that through time, access to resources and elevated social status have been very important in determining the reproductive success of males. Respect of peers is a major determinant of social status, and considering the ever-present hidden agenda that evolutionary history has provided, it is not at all ironic that the proximate goal of the participants in these altercations is to demonstrate that they "have balls."

Rape is a special form of violence that men inflict on women. It is not unusual for women in our present society to see rape as misogyny, an expression of hatred of women. A victim of rape who recently chose to make her story public was incredulous that as her attacker left he told her how pretty she was. The columnist who reported this story, also a woman, was equally shocked at the rapist's remark. Their bewilderment followed from their beliefs about the causes of rape.

Much sociological and psychiatric effort has gone into the study of rape, and the effort has illuminated some aspects of proximate cause. A smaller number of analyses have tried to put rape into an evolutionary perspective. Women react very negatively to rape for sound evolutionary reasons; the act not only subjects them to physical harm, it deprives them of their natural role of choosing with whom and when they will mate. Quite simply, rape is not in the interests of either women or their genes. As a further complication, rape can also compromise a woman's relationship with her husband; witness the tendency of men to lay blame on the victims of rape.

An evolutionary perspective suggests that, as in animals (e.g., ducks), rape has something to do with improving the rapist's Darwinian fitness.[30] In our mildly polygynous species we might predict that rape is most frequently performed by men whose educational level is low and whose economic (and thus reproductive) prospects are poor. This is in fact the case, not just in the racially troubled cities of the United States, but in the ethnic and cultural homogeneity of Denmark, and in cultures where the marriage prospects of men are constrained by the need to purchase wives, the custom of "bride-price." Economics is only one ingredient, for the typical rapist is not only a young man but one that has low self esteem and little sense of purpose in life. Whatever the proximate cause of his psychological profile, his prospects for increasing his fitness through orthodox means are poor, and he usually knows it.

The victims of rape are characteristically young women early in their reproductive years. If rape were an expression of hate of women, this would not necessarily be so. In fact, the age distribution of rape victims is skewed to younger years than the distribution of female victims of murder.

This overview of rape should not be taken to mean that men high on the socioeconomic ladder never try to coerce women into granting sexual favors. The way many men behave may be determined by their power over women and by their perceptions of the risks they are running. Groups of young men in fraternities and athletic teams can behave in insensitive, exploitative, and even criminal ways toward young women. And through history women have routinely suffered rape from conquering armies. None of this, however, detracts from the force of an evolutionary perspective.

The incidence of infanticide among mammals was mentioned previously as an example of how its explanation was muddled by the confusion of individual and group selection. Infanticide is not unknown in human societies either. Consider the advice of Moses to his people following their victory over the Midianites, in which

> they slew all the males. . . . Now therefore kill every male among the little ones, and kill every woman that hath known man by lying with him. But all the women children, that hath not known a man by lying with him, keep alive for yourselves.
>
> Numbers, 31

What could be a more explicit set of instructions both for eliminating reproductive competition, present and future, as well as for assuring paternity among the appropriated females?

Infanticide does not require the conditions of war. It has been encountered by anthropologists in a variety of cultures, where it is almost always an expedient to address one of several situations.[31] One is the birth of a deformed infant. A second is the wrong father. A third, and numerically the most important, occurs when the mother, usually quite young, does not have the social support and resources to care for the offspring. Whatever moral judgment one cares to make from the comfort and security of a modern Western culture, these are outcomes that are evolutionarily in the interests of the mother's ultimate fitness.

Sometimes there is a selective killing (or neglect) of daughters, and there is some evidence that this is most prevalent in highly stratified societies. A likely explanation in evolutionary terms is that in such societies the reproductive potential of sons of the upper classes can be enormous, much greater than that of daughters, and amplified if the society is polygynous.

Infanticide also occurs in modern Western cultures, although the mechanism may frequently be neglect rather than a single decisive act. The frequency is low—less than three dozen per million children per age class per year—but the circumstances are interesting. Infanticide at the hands of the mother usually results from the same situations as in "primitive" cultures, and young mothers are more likely to be involved. Children who are at risk from parents are at much higher risk during the first year of life and are at higher risk from stepparents than from natural parents. The fables of Cinderella and of Hansel and Gretel are rooted in fact.

The relationships between men and women are like an intricately cut diamond whose appearance changes when viewed from different directions. In this chapter we have been peering at our reflections from just a few of its many facets, and we have seen that there is more to human nature than can be understood without biology. The social sciences can describe, but in their present state they are unable to explain, some of the deepest questions that are posed by the behavior of their subjects.

Let me now head off one reaction that is probably inevitable in today's social climate. This is not a political essay. In invoking evolutionary biology and the concept of ultimate causation I trust I will not be saddled with the view that because something is "biological" it is *necessary* or *appropriate* or *right* for human society or that I am defending any social or economic status quo. Quite the contrary, there are a number of aspects of human behavior—regardless of what their origins may be—that may be maladaptive or culturally inappropriate in the technologically complicated world in which we now live. Unarguably, homicide, rape, and a host of other forms of violence and exploitation are deplorable, yet despite both moral and legal sanctions they remain disturbingly ubiquitous. I have argued above, and I shall argue repeatedly again, that in order to address biological and social problems we must accept the inherent complexity of what is meant by the word "cause." In short, where there is a problem, there is much to be said for trying to understand it before attempting to solve it.

Parable or reality?

On August 23, 1989, a sixteen-year-old youth named Yusuf Hawkins went with two friends to the Bensonhurst section of Brooklyn in New York City to answer an ad for a used blue Pontiac automobile that one of the boys had seen. Yusuf and his friends were black, and unbeknownst to them they were entering an ethnic Italian neighborhood where some of the young men had worked themselves into a dangerous state of mind. The outsiders were met by a mob that has been estimated to have been as large as forty, and before he knew what was happening, young Yusuf lay dead of gunshot wounds.

This tragedy is easily seen as simply one more example of wanton urban violence and racism. It is all of that, but it is also more. The residents who set upon the three black youths were young and male. They were from the local community, and their victims were not. Furthermore, being black, Yusuf and his friends were easily marked as outsiders. Why had the mob assembled, looking for trouble from other young men entering the neighborhood?

The answer to this question is fascinating. Equally interesting, the answer has been widely reported with absolutely no discussion of its deeper meaning. One of the young Bensonhurst men had had an altercation with a young woman in the neighborhood, a onetime girlfriend who had black and Puerto Rican friends from outside the immediate community. The young man took exception to this display of independence, and on the day of the murder the two of them had exchanged unpleasantries. According to newspaper accounts of the testimony at his trial, he insulted her, and she taunted him with the threat of reprisal at the hands of her friends. Apparently he took her seriously, and he seemingly had no trouble in persuading a sizable number of compatriots to defend the local honor, turf, and male prerogatives from outside invasion. When Yusuf and his two friends appeared later in the day they were promptly set upon, and Yusuf was killed.

The inability of young men to control the behavior of a young woman of shared ethnicity when she showed an interest in men from outside the group was clearly one of the ingredients of this tragedy. Lest the reader dismiss this as gratuitous hypothesizing, know also that in the subsequent criminal trial, one of the defense attorneys did his best to blame the entire incident on the young woman, a tactic he obviously expected might resonate with some of the jurors. Moreover, some residents of Bensonhurst offered the opinion that the woman was responsible for the killing, and she reported having been threatened. At a number of levels in our culture today, there is a deep acceptance of men's efforts to control the behavior—read reproductive lives—of women.

The story continues, for the responses of members of the community to the murder and subsequent trials are equally revealing. There was considerable group solidarity in the aftermath of the killing. Despite the number of residents who were on the scene, the prosecution had great trouble finding anyone willing to testify as a witness, partly due to fear of retaliation by other residents. "Bensonhurst amnesia" it was called. Bensonhurst nepotism would have been equally apt. There was concern for the image of the group. Following a conviction of one of the young men, a resident was quoted as saying, "The system condemned Bensonhurst yesterday, and today it vindicated Bensonhurst." Another asked, "Why should everyone suffer because of the action of a few?" Residents also found a material basis for concern, complaining that their property values and businesses were suffering from the publicity. The issue was cast in a manner calculated to downplay its significance: " . . . it was a freak thing. They were young, and they didn't know what they were doing." Or it was described in terms that were supposed to sound understandable and thus morally acceptable, in the same vein as the legal defense that had been based on the suggested culpability of the woman: "Residents insisted that the killing was a mistake provoked not by racism but by a desire to defend home turf." Community expression also took on a taunting, derisive tone with displays of watermelons when the residents witnessed demonstrations by black citizens from other parts of the city, responses subsequently attributed to "outsiders."

There is much of interest in this ugly story, and it gives us another invitation to contemplate causes, immediate and remote, proximate and ultimate. Some of its essential elements sound familiar bells. The Trojan War? The hatred of Montague for Capulet? These themes recur in literature because they recur in life. Yet there was nothing even faintly romantic about the loathsome behavior of the young men of Bensonhurst. How do we explain what happened? Some blame individuals, others society; but why should we assume that these are alternatives? What do we really mean by human nature? This question brings us without further ado to the interplay of proximate and ultimate cause.

5

Getting from Genes to Behavior

> Although it is self evident that, while each step in development is only rendered possible by the preceding steps, the whole course of development is nevertheless ruled and guided by the essential nature of the future organism. . . .
>
> K.E. von Baer, 1828[1]

Instinct and the myth of "Biological Determinism"

It is often said that there is no more sterile exercise than attempting to attribute particular behaviors to either heredity or environment. The effort is indeed without point or merit for most of the behavior of vertebrates. What is astonishing, however, is the frequency with which the presumption of which "either nature or nurture" recurs and the forms it takes. Virtually every account of sociobiology in "science" sections of the popular press is reduced to this simplistic notion, but one does not have to search far to find serious scholars who have unwittingly garroted themselves on the same clothesline. Thus we find an eminent anthropologist writing (in criticism of sociobiology and E.O. Wilson):

> The notion of the genetic prescription of behavior to which Wilson appeals is precise and understandable. It refers to the phenomenon of behavior being directly determined by genetically coded information, as in the case of the mosquito with its closed sequences of rigid behaviors programmed by the genes. There is, however, no comparably infrangible genetic prescription of the observable range of human behavior. . . . [2]

What then of genes and more "open" programs in which all of the steps in a behavioral sequence are not prescribed?

> . . . it is precisely because of [the] marked human capacity for non-genetically determined alternative action that sociobiological

70

theory, when applied to human populations, is irredeemably deficient.

The writer would have us understand that the existence of alternative behavioral choices, to be exercised with the benefit of learning, places the behavioral program beyond the reach of natural selection and therefore outside the concern of evolutionary biologists. That is what he seems to mean, and yet he also acknowledges a role for evolution when he writes:

> Thus, while it is certain that [learning and memory] which are the essential prerequisites for an open program of behavior have been evolved by natural selection, it is equally true that they are also mechanisms which, in their operation, do *not* directly involve the genetic code.

Before we proceed further, we had better examine this last statement closely, because it contains a serious confusion. What are the relationships of "closed" and "open" behaviors to the genes? The writer, I submit, has it precisely backward. In explaining why, I am hard pressed to improve on Richard Dawkins's brief but eloquent description:

> The reason why [genes] cannot manipulate our puppet strings directly [has to do with] . . . time-lags. Genes work by controlling protein synthesis. This is a powerful way of manipulating the world, but it is slow. It takes months of patiently pulling protein strings to build an embryo. The whole point about behavior, on the other hand, is that it is fast. It works on a time-scale not of months but of seconds and fractions of seconds. Something happens in the world, an owl flashes overhead, a rustle in the long grass betrays prey, and in milliseconds nervous systems crackle into action, muscles leap, and someone's life is saved—or lost. [3]

The execution of a behavior thus involves the operation of neural circuits that were laid down in development. The mouse's attempt to escape, an example of a relatively closed program like the blink of an eye or the jerk of a knee, does not require the construction of new circuits. It needs only their operation, which takes place on a time frame of small fractions of a single second, and it therefore does "*not* directly involve the genetic code." On the other hand, if the open programs of behavior involve learning, the time frame changes, and there is much reason to believe that the processes that occur in the nervous system include the synthesis of new proteins. But I am getting ahead of myself. The immediate point is that by contrast with simple reflexes and other relatively invariant or closed programs, in its *operation* an open program of behavior is very *likely* to involve the participation of the genetic code.

Finally, the heart of the matter is put before us:

> The issue at stake is the extent to which human cultures, and the behaviors which are part of them, can be accounted for by genetic determinism.[4]

At little risk of oversimplification, this kind of thinking can be summarized as follows. "Genetic" and "biological determinism" have come to be code words for forms of behavior that unfold along a fixed path and cannot be significantly altered by environmental contingencies. Any deflection of behavior that seems to involve choice (for human behavior, read "conscious choice") must therefore be something else. As this "something else" is, by definition, not genetic, it should, according to this fallacious argument, lie beyond the scope of evolutionary biology.

The roots of this dichotomous view of behavior run deep, for "biological determinism" is just instinct warmed over. A number of years ago the psychologist Frank Beach[5] pointed out that from the Middle Ages until the nineteenth century, instinct was a theological rather than a scientific concept and referred to the gamut of apparently purposeful (adaptive in post–Darwinian language) behaviors exhibited by animals. It stood in contrast to human behavior, which was motivated by reason. Instinct was a logically necessary construct, because the exercise of reason was the path to the soul's salvation, and as only humans had souls, presumably only humans could reason. In the nineteenth century the same sort of binary classification was extended to scientific usage, with instinct coming to mean the alternative to learned behavior. As Beach recognized, there is no theoretical justification for supposing that behavior must be either genetically programmed or acquired entirely by experience. And, in fact, such a view of behavior is quite simply wrong. Moreover, as Beach pointed out, a classification in which instinct is defined as something it is not is operationally unsatisfactory. Logically no behavior should be classified as an instinct unless it is first shown by observation and experiment to appear without any contributions from learning, a consideration that is frequently ignored and is extremely difficult to deal with experimentally.

Today it is recognized that behavioral phenotypes are the result of the interplay between internal (genetic) and external (experiential or environmental) factors. Consequently the concept of "species-specific behavior" I find more useful than the over-burdened term "instinct." Species-specific behavior is exhibited by most members of the same species of the same sex and age and under equivalent circumstances. It is generally adaptive. It is not synonymous with the common understanding of instinct.

There are many examples of species-specific behavior that can be drawn from the literature of ethology, but I shall mention here only two, selected because they illustrate the interplay of internal and external factors in the development of vertebrate behavior. A few hours after hatching, goslings start to follow the mother goose when she moves away from the nest. This behavior is also characteristic of a number of birds that nest on the ground, such as ducks and many shorebirds. Moreover, it is adaptive in that it keeps the large clutch of down-covered, mobile hatchlings close to the protection of a parent. If the adult bird is not present, the goslings will follow any moving object, even a human being, and subsequently behave toward that object as though it were the mother. This seemingly unreinforced, single-trial learning is called *imprinting*.

Note that the genetic program for this following behavior is not complete, an important detail being supplied by the environment after hatching. In other words, the genetic program does not equip a gosling to recognize a mother goose from all other objects in the world; rather it creates a short time in the developmental process—a *critical period*—in which the gosling's central nervous system is primed to receive and store some rather specific sensory information about mother's identity. Moreover, the information received at the critical period is important in determining the subsequent behavior of the young bird. The process works with high reliability in nature because it is rare that the mother bird is not the object of imprinting. The intervention of an ethologist is not an event that has influenced the evolutionary history of geese and ducks, and under normal circumstances their natural history virtually assures a normal outcome.

Another example that illustrates both the interaction between genetic and environmental factors and the notion of critical or sensitive periods in the development of behavior comes from studies of the ontogeny of bird song. The work on chaffinches and white crowned sparrows has been particularly instructive.[6] If these songbirds are reared in isolation, their adult vocalizations are not completely normal. If the birds are deafened at hatching, so that they can hear neither themselves nor other birds, their adult songs are even more distorted. The songs of deafened birds, however, are not completely random. It is as though the birds possess some kind of internal representation of their species' song against which they compare and refine their own vocalizations. Deafened birds, unable to make the comparison, nevertheless produce a song that has a number of the appropriate elements present, and a match to the template is improved if the birds are able to hear the songs of other individuals during a sensitive period during the first months of life.

For these species there are severe limits to the degree to which the adult song can be made to differ from the species-characteristic form. Some variation can be introduced by exposing the birds to different sounds during the early sensitive period, but the internal template limits the amount of novelty that can be learned. The amount of flexibility is sufficiently great, however, that different populations of wild birds occupying different geographical areas may sing recognizably different "dialects."

In summary, these two examples of the development of species-specific behavior illustrate several features of general importance. First, the development of behavior involves an interplay of genetic and environmental factors. There is something in the genetic code that makes a bird a chaffinch and not a chicken, and part of being a chaffinch is singing a chaffinch's song. Likewise, part of being a gosling is tagging along after mom. But in the process of development, the bird's song comes into full being partly as a result of listening and practicing, and for the gosling, knowing which object is mother requires that the nervous system be primed with sensory input. Second, the interplay of genetic and environmental influences is not random in time, for there exist prescribed intervals during which the developing animal is particularly susceptible to specific external influences, and it is at these times that the developmental process can be tricked. Finally, the environmental influences are not random in character, for the developing animal may be much more susceptible to some external events than others. This last point is another way of saying that not everything can be learned with equal facility. This is an additional concept that we shall deal with in due course; but first, a few more words about development.

The ontogeny of behavior follows general principles of development

Behavior—what animals do—depends on their nervous systems. More precisely, behavior is determined by the microarchitecture of the nervous system—by an enormous number of specific functional connections (called synapses) between nerve cells (neurons) that have different shapes and that communicate with one another using different chemical messengers (neurotransmitters). The development of behavior therefore becomes, at one level of analysis, a component of one of the most central, difficult, and elusive problems in all of experimental biology: How does a fertilized egg, a single cell, give rise to an entire functioning organism with many different kinds of cells each one of which performs a different task? The code for this process clearly resides in the genome: Moss genes produce mosses and mouse genes produce mice. But just what information is explicitly stated by the DNA? Does the genome specify, in detail, all of the connections a developing nervous system makes within itself? A simple calculation shows that this is not possible. The human brain is estimated to contain about 10^{12} neurons and roughly 10^{15} synapses, but human chromosomes contain about 10^5 genes. Even if these estimates are off by one or two orders of magnitude, one can see that the instructions for wiring together the brain must be quite general in character. There is simply not enough information in the genetic code to specify in advance every synaptic connection, let alone the finer details of neuron geometry.

Biologists have recognized for decades that the process of development involves an interplay between information coded in the genome (genetic factors) and a continuum of external signals influencing how that information is expressed (*epigenetic* processes). In response to these signals, local populations of growing and differentiating cells change their character, frequently irreversibly. A central feature of this interplay is that each stage in the process creates the conditions necessary for subsequent steps to occur. Although this feature of development has been recognized by experimental embryologists throughout the century, until recently few developmental biologists have been interested in the ontogeny of behavior or of human mental function, and relatively few psychologists have immersed themselves in developmental biology. Consequently, with a few notable exceptions, the conceptual framework that I present in this section has been slow to enter psychology, as described by Ronald Oppenheim[7] in an interesting essay on the history of epigenesis and preformation (the alternative and now discarded idea that the fertilized egg contains a miniature analog of every adult structure) as guiding ideas in the ontogeny of behavior.

The principle of epigenesis is at work in the very earliest stages of differentiation. For example, by the third week after fertilization, the embryo begins to form a nervous system. A group of outermost (ectodermal) cells along the dorsal midline are induced to become the future nervous system through the influence of a chemical (a peptide) that is produced by underlying (mesodermal) cells. "Induced" is a splendid descriptive word for what happens, and "induction" is actually the technical term that developmental biologists call the process. The presence of the chemical inducer sets the overlying ectodermal cells on a course of differentiation that is different from cells in neighboring regions of the embryo. Individually, the cells start to become either neurons or the associated supporting cells called glia. Collectively they form a hollow tube that in time changes further to become the spinal cord and brain. Moreover, the fates of these neuroectoderm cells become specified sequentially along the anterior-posterior axis of the developing embryo, presumably by a gradient of the inducing factor. The molecular details of how all of this occurs are not yet clear, but the process is believed to involve the local expression of different genes whose function is to regulate the expression of still other genes.[8] Thus among the first of the external signals influencing the developing nervous system is a chemical substance, produced elsewhere in the embryo, whose qualitative effects are in part determined by the spatial and temporal relations that the target cells have to one another in the embryo. The process is very complicated, because the embryonic cells are themselves moving, individually migrating and specializing, and collectively molding the shape of the embryo.

Subsequent differentiation of the nervous system continues to involve cell divisions and the creation of identifiable classes of cells, migrations of various classes of cells to new locations, growth of the long extensions of nerve cells called axons and shorter processes called dendrites, and the formation of specific synaptic connections between axons and dendrites from different cells. Furthermore, the various functional classes of cells appear at different times in development in response to specific local conditions. Differentiation continues to involve the activation of select subsets of genes. For example, the fertilized egg contains the information to direct the synthesis of all of the synaptic transmitters found in the adult; therefore, when a class of nerve cells commits itself to the use of, say, serotonin, genes for the appropriate synthetic enzymes are selectively activated and genes for other transmitters lie quiescent. Molecular signals may influence many cells or potential cell types, as in the example of neural induction, or they may have more restricted targets, as for example the influence of "nerve growth factor" on the induction of enzymes needed by neurons that use the transmitter epinephrine.

The identification of the external signals that control or modulate the development of the nervous system is in a rudimentary state. What seems clear, however, is that there is probably a variety of influences of rather different kinds. We have mentioned diffusible molecules, which are one example. The growth of the long cell processes called axons that make up the fibers in nerves is frequently directed along a scaffolding of supporting glial cells, or even other axons, indicating the importance of surface contacts. In other words, growing axons seem to "feel" their way along, detecting molecular cues as they go. Moreover, some of these contacts involve very specific recognition mechanisms, evident as cells grow into distant regions and select appropriate partners with which to synapse.

The process of development also includes elements of chance. For example, many more neurons of the type that control muscles (motor nerves) are formed than are ultimately employed, and those that fail to make the right connections in timely fashion subsequently die. There is thus a programmed redundancy in the pool of cells whose developmental fate is to be a motor neuron, and if a cell is not needed when all of the necessary synapses are formed, it is not kept.

Furthermore, the influences of one cell upon another can have elements of reciprocity. The formation of contacts between the terminals of a growing motor nerve and the muscle it is destined to activate makes the muscle less receptive to other nerve fibers and causes receptor sites for the synaptic transmitter on the muscle membrane to congregate in the region of the synaptic junction. Moreover, the speed of contraction of the adult muscle—the chemical character of the contractile machinery—is influenced both by the frequency at which the motor nerve delivers nerve impulses as well as by an unidentified trophic molecule secreted from the nerve terminal.

To summarize, although the molecular details of the process are far from clear, the nervous system develops through a complex series of genetic and epigenetic events in which the results of each step set the stage for the next. Along the way, cells, in effect, make a series of developmental choices, for the most part irreversible, about which genes to activate, in which directions to grow, and with which neighbors to settle. The outcome of each choice is determined by both the immediate local environment and the past history of the cell. The time at which each external influence plays its role on a given cell is thus crucial; the process of differentiation is a stream of critical periods.[9]

Let us look at several more specific examples that extend to times later in development, for they enlarge the concept of external influences. Although the sex of mammals is genetically determined, the process by which this determination occurs is such that the genome can be fooled.[10] Genetically, males differ from females in one of the pairs of chromosomes. In females these two sex chromosomes (called X chromosomes) are equal in size. Males, however, have one X chromosome and a small Y chromosome. A mammalian egg always has a single X chromosome, whereas sperm have either an X or a Y chromosome. The sex of offspring is therefore determined by the sperm, a particular irony considering the frequency with which, throughout history, wives have been blamed for the inability to produce sons. (In birds, the roles are reversed, and it is the genotype of the egg that determines the sex of the offspring.)

Maleness of mammals results from the production of testosterone during critical periods during development and the action of testosterone (or one of its derivatives) on the developing embryo. The earliest developmental fork in the road to maleness or femaleness occurs when a gene on the Y chromosome becomes activated and sets the fetal gonad to becoming testes instead of ovaries. At this writing the product for which this gene codes has not been identified, but it is thought to be a protein that binds to DNA and regulates the activation (the transcription) of other coding regions in the DNA. The developing testes begin to produce the steroid hormone testosterone, which, as we shall soon see, has, in its turn, pronounced effects on the subsequent development of the rest of the reproductive system and on the brain.

In humans the early activation of the sex-determining region on the Y chromosome is thought to occur somewhere between the forty-third and forty-ninth day of development, and herein lies a digressive tale. Shortly after the discovery of this gene was reported, it received some publicity in the popular press, and under "Religion Notes" *The New York Times* gleefully proclaimed "Talmudic wisdom confirmed. . . . The Talmudists cited page 60a in the tractate B'rachot in which the ancient rabbis question whether it is worthwhile for the husband of a pregnant woman to pray for a son. The rabbis determine that such prayers are worthwhile up to the 40th day of pregnancy, but after that point the supplication is a 'vain prayer' since the sex is already determined."[11] We can speculate that ancient rabbis were not operating totally in the dark; they had probably taken a close look at six- to eight-week-old fetuses and had taken their cue from what they saw. The religion editor of *The New York Times* and his Talmudist sources, however, have missed the point. In the course of normal development, the sex of the offspring is of course placed beyond the reach of prayer at conception, not when the sex-determining gene on the Y chromosome is first activated.

As fetal development progresses, testosterone synthesized by the developing testes becomes a major player in organizing the rest of the reproductive system. An early effect of testosterone is to promote the continued development of the Wolffian ducts into the male sex organs, a process that is accompanied by degeneration of the Müllerian ducts, which is in turn stimulated by another molecule (a protein rather than a steroid) secreted by the developing testes. Without this organizing influence, the Müllerian ducts continue developing to become the female reproductive organs, and it is the Wolffian ducts that degenerate. In humans there is a genetic mutation that blocks the molecular target for the androgen (a generic name for male steroid hormones), and such individuals, although genetically XY, have the external genitalia and sexual interests of females.[12] They are not reproductively functional females, however, because they do not develop the internal female reproductive organs. They have testes, which do not descend and do not produce spermatozoa as in a normal male. Conversely, genetic females that receive an early exposure to androgens develop the external genitalia of males. This can result from a genetic defect in which the adrenal glands produce the wrong steroid hormone, one with the activity of testosterone. From such observations as these it is apparent that in mammals the female condition is in a sense the base condition, and maleness results from a supplemental occurrence during development. This makes for an engaging contrast with the biblical story of Eve's creation from Adam's rib.

The steroid hormones also influence the developing brain. In rats there is a critical period during the first five days after birth, which fact has allowed considerable experimental manipulation. Genetically male rats that are castrated on the day after birth do not receive this critical exposure to androgen, and if they are subsequently given estrogen as adults, they exhibit female behavior. Exposure of the developing rat to androgen during the critical period not only determines aspects of adult behavior (e.g., the receptive posture called lordosis in females, more aggression in males), but produces a demonstrable sexual dimorphism in other features of the brain. In terms of neuroanatomy, differences have been described in the preoptic area of the hypothalamus at the level of cell number[13] and in the distribution of synapses on dendritic shafts relative to spines.[14] These small differences in architecture in the regions of cells near synaptic junctions with other cells can be seen with the electron microscope, but their precise functional significance is not yet known. They very likely correlate with functional differences, however, because all neural activity has a structural basis, even if it is only expressed at the level of molecules. At a biochemical level, hypothalamic cells (a part of the brain) of adult males show lower binding of estrogens to intracellular receptor proteins, and the male brain does not exhibit a cyclic pattern of secretion of luteinizing hormone from the anterior pituitary, whereas the female does.

Sex differences in behavior are not limited to the act of reproduction, and the sexually dimorphic changes in neuronal architecture that result from early differences in the hormonal environment extend to the cerebral cortex. Male rhesus monkeys show an earlier specialization of the prefrontal cortex in certain spatial learning tasks, and prenatal exposure of female fetuses to androgens will abolish this difference.[15] In humans the left-right specialization of the hemispheres develops later in girls than in boys,[16] although there is no direct experimental evidence that this is the result of differences in the prenatal environment.

The example of sex determination is interesting for two reasons. First, it shows that external influences on the genome of developing cells need not originate from adjacent structures. Hormones released by remote tissues can reach throughout the body of even an adult, and during development they can affect the transcription of specific portions of the DNA in select target cells. (The next example, however, will expand even further the concept of external influences.) Second, it provides a clear instance of measurable differences in the structure and biochemistry of brain cells that correlate with differences in behavior. There is really no alternative to the premise that structural and biochemical properties of neurons and groups of neurons lie at the basis of behavior, but there are relatively few natural examples in which variations in mammalian behavior can be correlated with structural properties of individual kinds of nerve cells. A principal reason is that the tools currently available are, in general, not adequate to the task.

Although there are clearly sex-related differences in the cellular structure of mammalian brains, the full implications of these findings for human behavior are far from clear. Humans retain an enormous behavioral plasticity, and the early social environment can doubtless influence the way men and women relate to one another. But at this juncture in the argument we have seen that *Homo sapiens* not only exhibits sexual dimorphism, we have also witnessed how some of this comes into being during development, how and why it is widely shared with other species, as well as how it is widely manifest in human behavior around the world. What we do with this information in order to produce a more equitable social structure is a different matter, and not one that is intractable. The challenge is not made any easier, however, by ignoring the reality that evolution has made us what we are.

Studies of the visual system have revealed still another dimension to the interaction between intrinsic and extrinsic factors during development.[17] Individual cells in the visual cortex of mammals respond to precisely oriented edges, or moving edges, in restricted parts of the visual field. Moreover, many of these cells receive convergent messages from corresponding parts of both retinas. These cells are properly "wired" at birth or shortly thereafter; that is, during development all of the neurons between the receptors in the retina, the lateral geniculate nucleus (a relay station in the thalamus), and the visual cortex have found the correct neighbors with which to synapse, and visual information is processed along this pathway much as it is in the adult. But something is still missing in order for the pathway to be consolidated: namely, sensory experience during a critical period shortly after birth. If mammals are prevented from seeing for weeks or months postpartum, they appear unable to resolve images and behave, as adults, as though functionally blind. If only one eye is covered, even with a translucent screen that allows light to pass, but no images, subsequent examination of the visual cortex reveals many fewer cells that respond to stimuli from both eyes. The synapses from cells driven by the covered eye fail to become validated and they decrease in number, their places having been taken by connections from the functional eye. A similar outcome occurs if, instead of covering one eye, the muscles that move the eyeball are cut so that the operated eye cannot be made to look at the same object as the unoperated eye. When the two eyes are unable to view the same point in space simultaneously during the critical period, they are subsequently unable to work together in analyzing the world. Again, this is because of a failure of nerve fibers with information from the two eyes to form synapses on the same cells in the part of the cerebral cortex to which they report. Thus we see that sensory experience—or the lack thereof—can cause irreversible structural changes in the nervous system and that this can have profound consequences for behavior.

The larger value of this discovery is that it gives us a perspective on other aspects of the ontogeny of behavior by extending the concept of critical or sensitive periods. The Harlows'[18] well-known study of the development of social behavior in monkeys fits this picture well. Young monkeys deprived of physical contact with both their mother and their peers grow up with severe behavioral deficiencies and invite the characterization of being neurotic. Somewhat less severe effects are produced if the infant has contact with its mother but not with other youngsters. The effects of deprivation are, furthermore, difficult to reverse. Similar stunting of the emotional growth of human children results from early lack of social contact[19] and the acquisition of first language seems to occur during a sensitive period.[20]

We should therefore view the processes of development as extending well after birth and involving periods of time in which certain broad classes of external events are necessary for fine-tuning the synaptic structure of the nervous system. For example, this is the role of play, so prominent in social species from primates to carnivores, where functioning as an adult involves complex interactions with other members of the species—interactions in which aggressive and cooperative impulses need to be appropriately balanced.

Many of the later stages in the maturation of the mammalian nervous system have been studied as part of psychology. They have therefore been described as learning or socialization, and generally in language that does not do much to stimulate thought about the underlying cellular processes. The work that has just been discussed, however, suggests that a more biological orientation would be useful, at the very least in providing a conceptual framework more likely to provide deep understanding in the future. There is, in short, good reason to see critical and sensitive periods in the development of behavior as part of a long developmental process in which the information contained in the genetic code can only be expressed through an intricate series of interactions of the partially completed product with a variety of extrinsic events. The final tuning of the nervous system—the construction, consolidation, and validation of all of the synaptic connections required for the full array of social behaviors—requires a variety of sensory and motor experiences extending, in our species, for years after birth.

Development clearly involves changes in structure, and what is ordinarily called learning must also involve structural changes in the nervous system, too. The immunity of long-term memory to changes in metabolic rate due to lowered body temperature or to anesthesia imply some structural base. Cooling or disruption of function with anesthetics do not erase memories. Memory is therefore not simply the result of a pattern of metabolic activity of neurons to be lost like the letter you were composing on your computer when the power failed. That memory survives when the brain is even partly shut down means that it must have a physical representation in the architecture of the brain. This structural basis is believed to reside in the organization of neurons and the synapses that connect them. Alterations of synaptic structure[21] or efficacy[22] as a result of sensory experience or learning have been described. The key to understanding learning and memory will likely be found by further study of structural changes at both the cellular and molecular levels.

As an aside, the tradition of viewing learned behavior as somehow nonbiological has its counterpart in medicine. Psychiatrists traditionally classified mental disorders as organic—meaning some relatively gross structural abnormality appears in postmortem examination—or functional—meaning no structural changes are evident. It is now more commonly recognized, however, that all mental disorders have a biological basis, whether there be gross tissue damage due to injury, genetically prone, environmentally provoked imbalances in transmitter biochemistry, or subtle morphological and biochemical abnormalities caused by inadequate sensory experience (social environment) at an early sensitive period.[23] Traditional diagnosis compounds the nominal fallacy— the illusion that by naming a phenomenon one has gained understanding. As with the supposed distinction between nature and nurture, the classification of mental disease as organic or functional makes understanding more difficult.

As much of learning occurs as a natural part of development and shares with the development of gross morphological change alterations in cellular and molecular structure, is it sensible to view all of learning as part of the process of development? The aphorism about the difficulties of teaching old dogs new tricks notwithstanding, mentally active people like to feel, with considerable justification, that they continue to learn throughout their adult life. But development, with its critical periods, would seem to be a series of opportunities, which, if once missed, are gone forever. Surely this is not a description of learning as it is generally experienced. Failure to optimize one's income tax return never prevents one from learning how to do the job better the next year. A useful way to view the relation between development and learning is through the common influence of the genes. Without question, the brains of adult mammals in general, and humans in particular, are endowed with a plasticity that enables them to continually adjust their behavior with experience. The developmental process does not tie down every conceivable synapse in a rigid and unalterable form, but leaves considerable scope for ongoing readjustment in the adult. At the same time it seems equally true that certain kinds of competence—perceptual, linguistic, social—do need to develop on schedule, or the deleterious consequences are reversed with difficulty, if at all. This is because the capacity for learning, like the development of body form, is subject to some genetic constraints.

The anthropologist Gregory Bateson, in a serious effort to place mind and biological evolution in a common framework, has contrasted ontogeny with natural selection.[24] Development is conservative and predictable; to use his terminology, it is a convergent sequence. Evolution, on the other hand, feeds on randomness. The next event is not precisely determined by the last and is therefore not predictable. The sequence is divergent. The formation of mind, and Bateson defines mind in sufficiently general terms that it is not necessarily a uniquely human attribute, is a stochastic affair. As with natural selection, the essence is exploration and change.

I believe this view of how brains work seriously underestimates the steering that evolutionary history imposes on the behavior of animals, including humans. In contrast, both proximate and ultimate aspects of cause are addressed if learning and development are seen in their proper relation to the genes. (Bateson in fact wavers and is not always consistent. Cultural transmission becomes a hybrid, and "there is, surely, *always*, a genetic contribution to all somatic events.") We will return to the evidence that learning is not a totally open process in a following section, but first a few final words about cellular development and change in the nervous system.

Species differ in the amount of external programming of behavior that is necessary during development. A caterpillar needs no practice to spin a cocoon, whereas a child needs considerable experience to master a language. Indeed, one might place animals along a scale according to how much external programming is required during development, and doubtless our species would be at one end of this array. Such a one-dimensional image of behavioral diversity would be misleading, however, because it speaks only to very general interspecific comparisons and does not recognize that in any one animal behaviors have diverse and interlocking ontogenies. Recall the example of social development in young primates that was discussed above. The process involves a mixture of behavioral elements developing at different rates with varying degrees of autonomy. Consider further the importance of nonverbal communication, and in particular, the smiling of humans. Smiling conveys information about emotions, and as it implies a state of pleasure it is frequently employed as a signal of reassurance in encounters between strangers. The fact that certain individuals can learn to use smiling deceitfully only underscores its general importance. But smiling is not something human infants must learn as a result of visual or auditory cues from other people; smiling first appears in congenitally deaf and blind children at the age of several months under conditions when they are apparently happy.[25] It does not take much imagination to envision the social difficulties an otherwise normal child would experience should he or she be afflicted with an inability to smile at all, or a compulsion to smile exclusively at inappropriate moments. This one defect, in a behavior that appears without any identified learning, would have manifold consequences cascading through the process of socialization. It would change the nature of the feedback loops that operate during the acquisition of social experience, and although one can anticipate compensatory behavior on the part of close friends and relatives, one can also imagine severe episodes of ridicule and ostracism and ultimately a warping of personality. The point of this hypothetical example, however, is not to generate disagreement about the details of the outcome, but to emphasize, from one more perspective, the essential inseparability of intrinsic and extrinsic factors in the development of mammalian behavior.

To summarize where we have come so far in this chapter, a detailed understanding of the proximate causes of behavior leads us, inevitably, backward along the paths of development. Some of these paths we find are much straighter that others, as though the genes had provided the zygote with an unambiguous road map. In other instances, however, we see that the developing animal received additional instructions along the way. Moreover, for complex behavior we find that different elements may have traveled along separate paths for part of the developmental journey. These paths may branch or merge, generating a tangled maze of social behaviors in which the origins of any path are hard to discern. For some of the straighter paths, instinct may remain a useful word. But for much of the behavior of higher vertebrates, attributing behavior to either learning or instinct is about as fruitful as arguing whether it is the sugar or the flour that makes the cake.

Lumsden and Wilson[26] have summarized the evidence that a prominent feature of the way the human brain deals with a complicated array of possibly conflicting information is to try to reduce the options to a dichotomous choice. This certainly is evident in the nature-nurture controversy, and understanding is not furthered by trying to allocate a percentage of human behavior that can be accounted for by the genes, with the balance to be attributed to culture. The biologist John Bonner has written, " . . . it is not clear that one will ever be able to determine to what extent any human action is genetically or culturally determined."[27] I would go farther and assert that the proposition has no meaning.

The other end of life: Why do we age and die?

To close this discussion of development, let's contemplate briefly the termination of the life cycle. Development from fertilized egg to adult is just one end of a series of biological changes that all organisms experience. For more than a century, development has represented a central challenge to experimental biologists, and it is currently a major focus of research activity. The inevitability of aging and death haunts the human consciousness and would seem to pose an equivalent intellectual challenge, yet we seem to have little in the way of explanation for why this happens to us. We plant annuals and perennials in the garden and we stand in awe of trees that have lived for centuries. We know that many insects may live for only days or weeks, a dog or cat for little more than a decade, while we and elephants and sea turtles may hope for several score years. But whether an individual is claimed by accident or infirmity, the end is inevitable. Why?

Sometimes the answer to this question is sought in terms of proximate cause. Perhaps our parts wear out, and if we just knew more about repair we could greatly extend our lives. And so we put a modest amount of money into research on aging, or gerontology, as the field is called. Interestingly, medical advances have increased the average life expectancy, but this has been accomplished solely by preventing death during younger years. Nothing in medical science has extended significantly the maximum age of about a century that individuals can reach; we simply increase the number of individuals who live to approach that maximum.

As we get along in years, a number of unpleasant things happen to us, including susceptibility to a host of ailments such as coronary disease, cancer, dementia, broken bones that do not heal, and so forth, and it becomes just a matter of time until one or another of these afflictions carries us away. We may not like to think about it this way, but we are programmed to age and die.

If that is the correct way to view the matter, then perhaps we can find some meaning, or at least explanation, if we wonder about ultimate cause. But to arrive at a sensible answer I think we have to recognize again the great variety of life-history strategies that living forms have adopted. Different forms of life cycles represent alternative ways genetic information has of propagating itself through time. It is not that one strategy is better than another; all that now exist have simply proved to be adequate to the challenges that they have met in their evolutionary history. Nothing will be gained by looking for the advantage of one compared to another, although we can make some sensible statements about ecological conditions that may either favor or compromise success of one or another strategy. To find the answer to our riddle, we should instead look for arguments that can be applied more generally.

It matters little where we start, so let's begin with ourselves. In our life history, our young take time to develop strength and experience, and it would be pointless for reproductive capacity to mature so early that novice parenting would not be successful. Within this constraint, imposed by the particular life history of our species, there is nevertheless reason for individuals to start reproducing as early as is consistent with a successful outcome. One reason why this is so is that the longer an individual postpones reproduction, the greater is the probability that he or she will either succumb to an accident before being able to reproduce or that there will be unwanted mutations in the germ line. If for any reason the probability of successful reproduction declines with age, regardless of physical condition, then those individuals that make an effort to reproduce early will, on average, leave more and healthier offspring than those that defer. In our species the risks of pregnancy are in fact greater for older women. Furthermore, earlier reproduction hastens the contribution of the bearer's genes into succeeding generations. Everything else being equal, selection will favor early reproduction, but not so early as to jeopardize success. In itself, this is a compromise.

Two other considerations bear on the argument. If the act of reproduction carries with it sufficient cost—for example, either direct risk to life or indirect risk through weakening or added exposure to danger—there may be an effective limit to an individual's capacity for reproduction. Perhaps it is for this reason that human females produce but several hundred eggs in their lifetime. If there is a limitation on total reproductive effort, and selection is encouraging early reproduction, the reproductive potential of individuals will decline after a certain age. As it diminishes, however, various forms of selection for somatic survival will be diminished. There will be decreasing need, in an evolutionary sense, to keep the body in the pristine shape it enjoyed when reproductive potential was high, and defenses against the ravages of tumors may relax. These changes may proceed slowly, as in humans, where an individual's contributions to inclusive fitness may continue through behavioral means long after direct capacity for reproduction has ended. Alternatively, changes leading to death may be cataclysmic, such as with salmon, where reproductive effort is compressed into a single act with no subsequent parental investment.

This argument is based on the observation that most genes have multiple phenotypic effects and the inference that genes may be beneficial at one stage of the life cycle and neutral or deleterious at another.[28] The process of development in fact involves the differential expression of genes, so in general genes can and do have different effects at different times. It is more speculative whether some genes and gene products that play a positive role in developing or in reproductively competent individuals may have deleterious effects later in the life cycle. Such genes would be selected for, but would contribute to the process of senescence. The uncertainty about the effects of estrogen supplements for postmenopausal women, however, provides an example that may make the concept seem more than plausible.

In summary, the notion that selection for continued integrity of the body will be relaxed in post-reproductive individuals is sound. On the other hand, the details of the cellular and subcellular compromises that are made during the life cycle, and their genetic substrate that is sifted by selection, will only become clear as we come to understand the molecular events that take place during development, maturation, and aging.

6
Evolutionary Perspectives on Volition, Learning, and Language

> The difference in mind between man and the higher animals, great as it is, certainly is one of degree and not of kind. We have seen that the senses and intuitions, the various emotions and faculties, such as love, memory, attention, curiosity, imitation, reason, etc., of which man boasts, may be found in an incipient, or even sometimes in a well-developed condition in lower animals.
> Charles Darwin[1]

How do we know that behavior evolves?

More than a century has passed since the publication of *The Origin of Species,* and the full implications of the Darwinian revolution are still being absorbed. The vast majority of persons who have acquainted themselves with the evidence understand that earth's great variety of plants and animals, including *Homo sapiens*, is the result of slow but ongoing processes of organic change. Evolution, as exhibited by morphological differences between species, is readily accepted by most educated people, if not through firsthand examination of the data then through a comfortable acceptance of the conclusions of those who have examined it.

There is therefore one answer to the question "how do we know that behavior evolves?" that is so obvious as to make the question sound trivial. Rats, monkeys, and people have different brains and different behaviors. Brains, like other morphological features, are the products of organic evolution. As brains generate behavior, behavior must also evolve.

For many the concept of behavioral evolution, at least of the human species, nevertheless remains a foreign, or pointless, or even odious idea. There is a variety of interlocking reasons why this is so. One reason is that behavior seems far removed from the linear sequence of nucleotides in the DNA of the nucleus, that string of molecular beads of which the genes are made and in whose sequential ordering resides all the genetic information an organism receives from its parents. Behavior is a property of the entire functioning organism, and the developmental pathways by which the information contained in genes is translated, supplemented, and embellished, finally to be manifest as behavior, are indeed complex, indirect, and hard to perceive.[2] In fact, sociobiologists have frequently confused their cause by alluding to hypothetical genes for complex behaviors such as altruism. Such genes are generally invoked in arguments that are spun to illustrate how natural selection might work to encourage the retention of a particular behavior in a population. These arguments have an allegorical quality, and as long as they are understood in that spirit, there is no problem. If interpreted literally, however, they can lead to mischief. There are not single genes for altruism and aggression, just as there are not single genes for arms or stomachs. Complex behavior, like complex morphology—of which it is one manifestation—does not map onto the genome in such a simple fashion. Our genetic endowment specifies a design for functioning and a strategy for living and being, neither of which can be expressed without the harmonious and sequential interplay of self with external things and events.

But that does not mean that behavior fails to lie close to the process of natural selection. Whether or not one animal reproduces where another fails can be decided by small differences in behavior. We should therefore expect behavior to be molded by natural selection. And in many cases it is. The fruit fly *Drosophila* is a favorite animal for genetic and evolutionary studies because it has a short generation time and is yet intricate enough (compared with bacteria, for example) to reveal biological principles that hold for many different kinds of eukaryotic organisms. In laboratory experiments in which stocks of fruit flies were subjected to artificially imposed selection, a variety of behavioral traits have been dramatically altered in ten generations or less.[3] These behaviors include attraction toward or away from light, a propensity to walk up or down with respect to the earth's gravity, mating success (which involves a courtship in which males vibrate their wings close to the females and "lick" the females' ovipositors), and selecting genetically similar individuals in mating. It is precisely because behavior can be the object of selection that biologists are justified in hypothesizing behavioral adaptations. In order for behavior to be altered by natural selection, components of the design and the strategy must, of necessity, be coded in the genes.

Some traits of organisms are constrained by evolutionary forces and are relatively invariant from individual to individual. For example, morphological structures that are very important for reproduction frequently exhibit more constancy than features that are less critical. The structures of flowers are therefore usually more useful than the sizes and shapes of leaves in distinguishing closely related species of plants. Some closely related kinds of insects can be most easily distinguished by their genitalia, which have become specialized and contribute to the genetic isolation of the species.

Some traits are conservative, change slowly with evolution, and remain with the lineage for long periods of time; others are more malleable and exhibit much greater variation. The relative rates of evolution of several categories of social behavior have been estimated by identifying the lowest taxonomic hierarchy within which there is discernible variation. The rationale for this approach is that a trait is evolving relatively rapidly if it shows great variation between species, but slowly if it is shared by all members of a more inclusive taxonomic group such as a family. A clear conclusion from this tabulation is that most of the traits that characterize the social lives of vertebrates—for example, group size, presence or absence of harem structure, presence or absence of territoriality, involvement of both sexes in rearing young—are not very conservative, exhibiting differences at the species or even the population level.[4] To anticipate the next chapter, such differences in behavior can reflect differences in genotype, but they frequently do not. There are many examples of behavioral differences that represent alternative phenotypic expressions of basically the same genotype.

Free will

Each in their own way, the Christian, Jewish, and Islamic religions all embrace the view that people have individual responsibilities to one another exercised through rational judgment. The view that

God holds man morally responsible for his actions is a way of recognizing the difficulty of having functional societies of any size if selfish and aggressive behavior directed toward other individuals were always to be explained, and thus excused, by appeal to a deterministic philosophy in which people's actions are totally controlled by an outside force. The concept of Satan is also an implicit recognition that the interests of individuals are never congruent and are frequently in conflict, thus the struggle to live a moral life and the need for encouragement and even reinforcement in order to win the struggle. We are told that we have the capacity and the duty to chose good and reject evil; by extension, we have free will.

A second impediment to understanding the role of evolution in human behavior is associated with this notion of free will. Evolution, like secular, humanistic philosophies, also calls attention to the emotional as well as the rational side of human nature. For some people, evolution raises the specter of that bugaboo genetic determinism, and determinism is by definition the antithesis of free will. At the risk of evoking a physiologically deterministic metaphor, the association of evolution and human behavior therefore has the capacity to raise hackles. How is it that we can speak of the evolution of human behavior without denying the existence of free will?

These concerns are based on dichotomous concepts as unrewarding as the view of behavior that posits either nature or nurture as its "cause." Brains are made of cells; cells are made of molecules. What brains do must therefore follow natural laws. Brains are physical devices, constructed from genetic blueprints crafted in evolutionary time and tuned and programmed through individual experience. Alternative behaviors are the product of this involved history as it interacts with immediate sensory experience, but there is no reason to suppose that this vast complexity places behavior beyond the reach, ultimately, of scientific explanation. If free will implies choice that is utterly decoupled from these formative events, it locates volition outside of science. It is the very complexity of behavior, however, that creates the illusion of complete freedom. The choices we make, both large and small, are so numerous, the full range of sensory experience so unpredictable, competing motivations so finely balanced, the vicissitudes of attention and memory so likely, the multiplicity of social interactions so fortuitous, and the array of causal factors both so large and so varied that behavioral responses can seem to be undetermined. In retrospect, however, one can frequently construct a plausible sequence of causal events, but like evolutionary change, chance plays a large role and predictions are at best contingent.

Because of this uncertainty, society works hard to channel the behavior of individuals, and the appeal through religion to particular norms of behavior is but one example of how the human brain becomes programmed to make certain kinds of choices. The cognitive capacity to weigh the pros and cons of the various alternatives that previous programming and present circumstances present, however, defines the scope of this freedom of will.

Evolution and learning

Both experimental work in the laboratory and comparative analyses of behavior observed in natural populations support the concept that selection and evolution can alter behavior, but we must continue to reconcile this concept with the observation that the behavior of individuals can be enormously diverse and variable. Specifically, how can learning—the acquisition of behavior during the lifetime of an individual—have anything to do with genes and evolution? Isn't learning so open-ended that it liberates us from the chains of our evolutionary past?

Although—as illustrated by the passage that introduced this chapter—one of Darwin's contributions was to perceive that behavior evolves, by the early 1900s the study of animal behavior, particularly in the United States, had become separated from the study of evolution and from questions of ultimate cause. Moreover, the evolutionary view explicit in Darwin's words was seen as both anthropomorphic and operationally unsatisfactory. As mental attributes such as joy or curiosity could not be observed directly, they were abandoned as important concepts in favor of a more objective approach in which animals were treated as "black boxes" whose observable behaviors were quantified in terms of measurable stimuli and the responses of the animals that the stimuli evoked. Sensible as this approach may seem, it led to the domination of psychology for many years by the school of behaviorism and a preoccupation with the rules by which a few kinds of laboratory animals learn. Today it is easy to be critical of the smothering effect of this tradition on progress in animal behavior, but it can be properly viewed only in historical perspective. Throughout the first half of the twentieth century the methods for asking questions about neural activity in the brain were so crude as to be virtually useless, so matters of internal workings were in fact best deferred. Nevertheless, the influence of behaviorism was so great in this country that a rekindling of interest in the evolutionary analysis of behavior ultimately came from Europe, starting with the ethologists Konrad Lorenz and Nikolaas Tinbergen.

There is in fact a minor paradox in this story. While eschewing evolutionary thinking about behavior, behaviorists proceeded to work on the assumption that all of the important rules about human learning could be deduced from the study of bar pressing by rats and key pecking by pigeons. By extension, Locke's view of the mind as a tabula rasa, a blank slate on which anything could be inscribed by appropriate experience, assumed particular prominence. To alter the metaphor, if the human mind at birth were an empty cup into which virtually any knowledge could be poured, how could we picture the mind of the rat from which we hope to learn about ourselves? Presumably it is a somewhat smaller cup.

Such images of the mind follow from the premise that evolution has worked only to adjust the amount of learning that an animal can exhibit. Or in other words, they rest on the assumption that it is simply the capacity to learn that has evolved. Can this position be sustained?

The psychologist Martin Seligman[5] has pointed out that animals seem to be arrayed on a dimension of "preparedness" for learning. Thus they may be prepared, unprepared (neutral), or counterprepared to associate specific kinds of sensory input with particular objects or events. An experimenter's success with any specific learning paradigm will therefore depend on the animal's preparedness, as molded by its evolutionary history. Imprinting, which we discussed previously, is an example of an association for which young geese are evolutionarily prepared, which is why the learning occurs with such ease and without the need for repeated reinforcement. The human nervous system is efficient in remembering facial features and associating large numbers of faces with particular individuals. We function as though we were designed for this task, and in this sense we are evolutionarily prepared. Most of us are much less able, however, to describe a face in sufficient detail to enable another person to identify the subject in a crowd of strangers. For that we are much less well prepared, a realization that emphasizes the subtle and still mysterious way in which our brains incorporate and utilize visual information about other human faces. This is an instance where the line (traditional with students) "I know it; I just can't explain it" actually has validity.

Other examples of relative preparedness for learning taken from the literature of psychology include the association by rats of taste with subsequent gastrointestinal distress, and pecking of illuminated keys by pigeons for food. By contrast, pigeons learn to peck keys to avoid shock only with great difficulty,[6] and it has proved very hard to reinforce yawning of dogs with food,[7] or licking or scratching by cats with release from a confined space.[8] Pigeons will learn auditory discriminations in a cardiac conditioning paradigm and visual discriminations in a key-pecking paradigm, but not vice versa.[9]

In general, animals seem prepared to make sensory associations if their responses reduce motivational states or physiological needs such as hunger or are a part of their natural behavior. For pigeons, learning to associate the peck of a key of a particular color with the delivery of food draws on the natural feeding behavior of this kind of bird. Visual discriminations and pecking are part of feeding, whereas auditory discriminations are not. For a grain-eating bird, visually directed pecking has little to do with avoiding immediate physical hurt of the kind that might come from a predator, whereas auditory stimuli, like the scream of a peregrine falcon, do. When examined in this light, the concept of evolutionary preparedness for learning seems to indicate some degree of adaptation. Put in an evolutionary context, natural selection has thus worked on both the number and the kinds of things an animal can learn. This phenomenon is also referred to as directedness of learning[10] or evolutionary constraints on learning.[11]

Psychologists interested in learning theory have paid less attention to these sorts of observations, preferring to focus on what they see as common to the process of learning in all animals (or at least the limited number of species that have traditionally received their attention). Only the "Garcia effect"[12] has stirred much interest, and it is instructive to examine this case more closely. When rats are made ill by heavy doses of X rays delivered after eating, they subsequently associate their delayed distress with the taste but not the shape or color of the food. This effect is unusual in the long delay that can exist between stimulus (food) and reinforcement (illness). By contrast, if rats are given a shock while eating, they remember visual or auditory cues, but not taste cues. Subsequently it was shown that birds can employ visual cues in long-delay food-avoidance learning.[13] An interpretation that is in harmony with both sets of observations is that eating-related cues are readily associated with subsequent internal distress, but which specific sensory modalities are involved depends on the species.[14] The adaptive significance of this interpretation is obvious.

Some psychologists have sought alternative explanations, usually under the rubric of differential salience of cues. Cues may have different salience for any of several reasons. An animal may learn to ignore certain cues in particular contexts simply because previous experience has shown them to be unreliable predictors. Or the motivational state of an animal—its current balance of drives— may determine which sensory input is heeded. Or simply the design of the experiment may determine the result. For example, the failure of taste-shock associations in Garcia's experiments have been attributed to the fact that the unconditioned stimulus (shock) is only effective if given after decay of the effect of the conditioned stimulus, which in the case of taste is long-lived.[15] This proposed mechanism does not seem to account for the long-delay aversion learning that can be visually mediated in birds, however. Similarly, although chicks can readily learn an aversion to red food, they learn a similar aversion to red water only if the experimental apparatus is arranged so that the birds must locate visually the spot from which to drink.[16] Thus the salience of red is determined by the nature of the response. In fact these authors concluded that in general the "examination of response characteristics is a good predictor of which cues are or will be selectively attended to in acquisition associations."

Although these words may sound like an argument for directed learning, their intended meaning is just the opposite. How is it that individuals from different research traditions in animal behavior can examine the same kinds of data and draw seemingly opposite conclusions? Part of the reason that learning theorists and evolutionary behaviorists tend to talk past one another has to do with the distinction between proximate and ultimate cause. For behaviorists, all of the relevant parameters of general process learning are, in principle, within the reach of the experimenter, either in the design of the learning task or through manipulation of the animal's previous experience. Therefore variations in performance between individuals, between species, or between sensory channels are, in the last analysis, all accounted for by differential salience of stimuli, which is an explanation of proximate cause. General process learning theory offers no overarching explanation for why species should differ from one another or why a given species attends to one kind of cue in one situation and another sensory input under other circumstances. Explanations of this zoological untidiness become, of necessity, post hoc appeals to salience.

Behavioral ecologists, on the other hand, who are more inclined to view behavior as broadly adaptive, see explanations in terms of ultimate cause. And to the extent that patterns of behavior are adaptive in the sense of adaptation described in Chapter 3, there can be little doubt that they are right. But because proximate and ultimate cause are but two sides of the same coin, does this schism really make any difference? I think it does. When the study of behavior is closed off from evolutionary considerations, it will neglect certain classes of explanatory hypothesis. Perhaps more important, learning theorists will fail to examine their underlying if unspoken assumption that the neural mechanisms that enable learning and memory have not themselves undergone evolutionary changes of important, qualitative kinds.

Let me illustrate with an example. Several years ago, in the course of some field studies of color vision of hummingbirds, I observed that although the birds could readily be trained to discriminate between feeders on the basis of color (wavelength) cues, they were unable to master discriminations when the cue was brightness (intensity of light).[17] Two features of this observation deserve note. First, this difference in learning behavior does not involve two different sensory modalities, as do many of the examples of preparedness for learning that were cited above. This greatly narrows the range of possible proximate causes that might be proposed in explanation. Second, for one familiar with the literature of sensory psychophysics, the result is counterintuitive. Brightness discriminations are generally easy for animals to make.

As the observation has been made repeatedly and under a variety of conditions, I believe it to be correct, and I can think of several possible explanations for the result. Perhaps the birds are unusual in having a very poor capacity for brightness discrimination and simply fail to process the sensory information. Considering their skill on maneuvering through vegetation in full flight without colliding with branches, however, this is hardly credible. Perhaps brightness has a low salience in selecting food because experience shows each individual bird that it is an unreliable predictor. This is more reasonable, for relative brightness is indeed an unreliable cue for hummingbirds to use in finding food in the natural world. I am skeptical of this explanation, however, because such learning-not-to-learn is generally a much weaker and more readily reversible effect than observed here.[18] Perhaps one can conservatively conclude that it is at least as likely that hummingbirds can much more readily learn to associate nectar sources with hue than with brightness cues. If this were correct, the reasons for a low salience of brightness would be built into the central nervous system as a result of evolutionary history and might be beyond the means of human experimenters to manipulate. Although experiments to distinguish between these hypotheses have not been done, my point is that the *nature* of the third hypothesis presents a challenge to the way general process learning is usually approached.

Biologists, while recognizing the universality of nucleic acids as genetic coding agents, have developed a healthy regard for the complexity of the mechanisms by which codes come to be translated into phenotypes during development. In a similar spirit, the cellular mechanisms by which brains make long-term records of experience may be importantly different in honeybees, lizards, rats, and humans. Or they may not. At this juncture, however, it seems only prudent to recognize that the general rules that describe learning in different kinds of animals may be based on a variety of kinds of physical changes in the brains of different species.

Communication and language

> And God said, Let us make man in our image, after our likeness: and let them have dominion over the fish of the sea and over the fowl of the air, and over the cattle, and over all the earth, and over every creeping thing that creepeth upon the earth.
>
> Genesis 1, 26.

These words express a deeply seated assumption about the place of *Homo sapiens* in nature that has influenced philosophical and political thought in the Western world for centuries. One of the cardinal features that separates man from the brutes, it is said, is the use of language, which is often taken to be the outward manifestation of conscious thought. It may be instructive, however, to consider language in a more extensive biological context.

Human language is a very elaborate form of communication, and systems of communication are widespread among animals, involving not only auditory but visual, tactile, chemical, and electrical sensory channels, sometimes operating separately and sometimes in combination. Signals associated with reproduction are particularly prevalent: Bright colors or other evidence of sexual selection are familiar in (usually) males of many vertebrates. Auditory signals include the songs of birds, the trilling and croaking of toads and frogs, and a variety of sounds made by mammals. Several families of fish generate weak rhythmic electrical signals that can announce the presence of specific individuals. Communication between members of the same species is particularly well developed in social animals, where information not only about reproductive status, fear, and anger are exchanged, but also a host of other matters arising from the interdependence of the individuals that make up the social system. A male rhesus monkey, for example, can convey information about where he belongs in the social hierarchy through his grooming behavior or even by his gait and posture. The richness of animal communication systems that has been documented in recent years was largely unsuspected a generation ago[19] and has rendered a number of earlier presumptions obsolete.[20] We shall consider a few of these that pertain to the uniqueness of human language.

Human language is symbolic in that elements of speech can bear a totally arbitrary relationship to the objects to which they refer. Moreover, language exhibits displacement, meaning that it can refer to objects or events spatially or temporally separated from the speaker. Because we frequently construct conscious mental images when engaging in this kind of communication, it becomes easy to assume that symbolism and displacement must be unique to human communication. Recent efforts to teach apes[21] or even a parrot[22] to communicate with arbitrary hand signals, objects, or vocalizations, however, make it clear that these animals have the cognitive ability to associate arbitrary gestures, objects, or sounds with actions or other unrelated objects to an extent that implies a true symbolic representation. What is far less clear, however, is whether the animals have any significant capacity to combine these elements of communication in ways that impart new meaning—that is, whether they can cope with syntax. As Donald Griffin[23] has pointed out, there is a danger in drawing definitive conclusions based only on rather artificial training situations, and more attention must be directed to the natural state. For example, vervet monkeys have three different alarm calls for snakes, leopards, and eagles, each of which elicits a distinct and appropriate response from other monkeys in the vicinity.[24] The calls convey not just fear, but specific information about the nature of the danger and the fact that different tactics are required for escape.

The social insects, challenged (in an evolutionary sense) by the task of coordinating the activities of enormous numbers of individuals in the same colony, have evolved some remarkable systems of communication that are supported by only several hundreds of thousands of neurons in the brain. As the "waggle dances" of honeybees are frequently referred to as the "dance language" of bees—but not without argument[25]—it will be helpful to examine the features of this communication system in more detail. When a foraging bee returns to the hive, she frequently conveys information to other workers about the food source she has just visited. In addition to the scent of the flowers that clings to her body, she communicates information about the direction, distance, and quality of her findings by performing on the surface of the comb a "dance" in which she moves forward about an inch while vigorously wagging her abdomen, then circles about and repeats the waggle run. Direction of the food source is indicated by the direction of the waggle run; distance, by the frequency with which the waggle run is repeated; and quality, by the vigor and the duration of the dance, and some acoustic signals.[26] There is no question that hive mates discover the nature of food sources through these dances and are recruited to fly to them.

There has been an objection to referring to these dances as language because the dancers are simply reenacting in miniature the outward flight path to the food. But the symbolic representation is more abstract than I have indicated. In general the dances are performed in the darkness of the hive and on the vertical surface of a comb. The worker bees that are receiving information remain in tactile contact with the dancer, but how is the direction to the food depicted with the dancer performing on a vertical surface? Amazingly, the outward flight direction is indicated relative to the horizontal angular position of the sun, but through a transposition of coordinates in which the current position of the sun outside the hive is represented during the dance as up. The angular orientation of the dance with respect to the vertical therefore represents the angle between the correct flight path and the sun as seen by a bee on the outward journey.

This communication system is usually thought of as being rigidly determined genetically. It is a very restricted system compared to human language, but in a number of respects it is finely tuned to prevailing circumstances. How a foraging bee is received upon her return to the hive depends on conditions in the hive. If food is plentiful there may be no dancing. If the hive is overheating, foragers may have difficulty finding bees to relieve them of nectar and pollen but no trouble in unloading water, which is hung in droplets and fanned for evaporative cooling. By their behavior, workers in the hive therefore can communicate something of the colony's specific needs and redirect the subsequent behavior of the foragers.

At times of swarming, in which part of the colony emigrates, the bees that leave must find a suitable new nest site. As in gathering food, scouts search and convey their findings by dancing. The form of the dance is the same as employed in foraging, but the information about quality refers to size and suitability of prospective nest cavities and is therefore entirely different. (Size is in some sense measured by the scouts, who pace off the dimensions of the new nest cavity, and quality is influenced by the nature of the entrance and the susceptibility to predation.)[27] The *meaning* of the dance therefore varies with the *circumstances* under which it occurs.

Most astonishing of all, individual scout bees that are reporting on possible nest sites do not dance their information like automata with no influence from the other bees. Although the details of the process are not fully understood, the swarm moves to its new nest cavity only when the scout bees have reached what is, in effect, a consensus, and their dances show the way to the same location.

The ontogeny of the communication behavior of honeybees has not been studied in great detail, but it is clear that it possesses some significant adaptive flexibility. During the warm months, individual bees live for several weeks. During this time they perform a sequence of jobs, starting with care of the larvae, progressing through other tasks such as the secretion of wax for combs, and ending their lives as foragers. The sequence is far from rigid, however, and can be modified to meet the immediate needs of the hive. For example, if the hive is separated from most of its foragers in a catastrophic fashion (for example, by a farmer closing and moving the hive in midday), young bees with no field experience accelerate their behavioral development and fill the breech.

It has been argued that the dances of bees do not compose a language, because a language must be culturally transmitted. The dialects of some bird songs, however, are culturally transmitted. The details of individual human languages are indeed specific to individual cultures, but those who argue that there is a "deep structure" to human language[28] are asserting the presence of a species-specific neural substrate, although they may not use those words. The acquisition of human language, like the songs of certain birds, therefore involves a mixture of genetic and epigenetic phenomena.

This brief discussion of communication carries three messages. First, it illustrates once again how the behavior of even an insect can be a complex of genetic and environmental influences, with behavior adjusted to contingencies in a manner that appears to our eyes both sensible and adaptive. Second, even the example of an insect, the honeybee, illustrates dramatically how social systems impose strong sources of natural selection for communication between individuals. Finally, although human language is indeed unique, it is not so for a number of reasons that have been offered in the past. Human language is quantitatively unique in the richness and diversity of its syntax. The implications of that difference are enormous; of that there can be no dispute. But given the vast difference in the sizes of the neural substrates with which the evolutionary process had to work, the communication systems of some of the social hymenoptera are perhaps no less remarkable than that of the most successful of the primates. We are part of nature, but we like to see ourselves otherwise.

7
Decisions, Decisions!

Reason [is] more often than not overpowered by non-rational human frailties—ambition, anxiety, status-seeking, fixed prejudices. Although the structure of human thought is based on logical procedure from premise to conclusion, it is not proof against the frailties and the passions.

Barbara Tuchman[1]

Drives and the evolution of the vertebrate brain

The single most important feature in the evolution of the mammalian brain has been the elaboration of new tissue—the neocortex—over phylogenetically more ancient structures—the limbic system and hypothalamus. The human cerebral cortex is a mantle of nerve cells that envelops the rest of the brain; it is what one would see if the skull were transparent. The neocortex made its first appearance as a relatively small bit of tissue in certain reptiles, but it is only in mammals, and particularly in primates, that it has become a major part of the brain. Interestingly, the neocortex, almost like hair, is a peculiarly mammalian innovation. Although birds, like mammals, have relatively large brains, their cortical tissue is not homologous to the neocortex of mammals, and the main visual and motor pathways have evolved in somewhat different ways. The cortex of birds has its mammalian counterpart in a structure called the corpus striatum or basal ganglia, which in mammals are primarily involved in organizing motor commands to the muscles. Whatever common functions are provided by the mammalian and the avian cortex therefore represent examples of convergent evolution.

The limbic system and hypothalamus are structures tucked away under the cortex, and in contrast they are evolutionarily conservative parts of the brain that have been around for far longer.

Homologous structures are present in reptiles and birds, where they subserve similar functions. The human brain was therefore not designed de novo; it is the result of a long process of evolutionary tinkering, and it functions through a frequently uneasy interplay between the hypothalamus and limbic system on the one hand and the neocortex on the other. It is desirable to understand something about the nature of this relationship in order to weigh properly the competing claims of "instinct," "free will," "self-awareness" and other emotionally burdened terms that are used to characterize behavior.

The hypothalamus and the overlying limbic system are involved in two general functions. The first is the regulation of the internal environment: the proper balance of salt and water, the control of blood pressure and temperature, and so forth. Second, these neural centers in the brain generate behaviors of varying complexity that are involved in eating, drinking, general arousal, aggression, escape from danger, copulation, and other activities. These behaviors generally also involve a set of motor nerves over which we have little voluntary control and which are referred to collectively as the autonomic nervous system. The autonomic nervous system regulates such activities as dilation and contraction of blood vessels, changes in heart rate, and contraction of the muscle lining the intestines.

The hypothalamus and limbic system are implicated in specific behaviors by some very direct and dramatic experiments. Under anesthesia it is possible to advance fine needle-shaped electrodes into various parts of the brain. The electrodes can be secured to sockets mounted in the skull, and because the nerve cells within the brain do not include any pain receptors, the animals can recover from the anesthesia and behave as normally as they did before the operation. A weak electric current applied through the electrode arouses nerve cells in the vicinity of the tip, and depending on where the electrode is located, the animal will exhibit, for example, eating or drinking behavior or manifestations of rage, fear, or pleasure. This is one of many kinds of evidence that local regions of the brain are specialized for particular functions, which demonstrates that homologous structures in different species share similar functions. Limited but comparable data on humans has been obtained by correlating behavioral changes with sites of certain kinds of epileptic foci.[2]

The literature of both psychology and ethology includes the concept of internal drives or motivational states associated with particular behaviors. Body needs and sensory stimuli create changes in the nervous systems of all mammals, including humans, and these changes can occur without the actual generation of a behavior such as feeding or copulation. Corresponding to these changes we recognize states of hunger and thirst as well as emotions of fear, rage, sexual desire, and pleasure. In fact, the behavioral homologies can be so evident that modest familiarity with individuals of another social species, dogs, gives many people confidence that they can tell when the animal is happy, playful, angry, fearful, curious, anxious or in some other state in which an adjective ordinarily descriptive of human emotions seems obviously appropriate to characterize their pet. This ability to "read the mind" of a dog is no doubt fostered by our mutual sociality; the fact that as species we have shared 10,000 years of close interaction, however, does not diminish the fact that our capacity to judge the other's moods, which is to some extent mutual, reflects some shared needs as well as similar neural processes.

The evolution of the neocortex poses a deeper mystery. The various sensory and motor functions that have been studied in single neurons of mammalian cortices seem to be carried out as well in lower centers of reptiles and amphibians. Furthermore, the cortex is not necessary for learning, because some capacity for learning seems to be a very general property of neural tissue. Birds and mammals have larger brains, relative to body size, than lower vertebrates, but the difference is not due simply to the addition of a cortex; all parts of the brain are correspondingly bigger.[3] What then does the cortex do? Although it would be naive to settle for a single answer to this question, one important idea is based on the observation that large parts of the cortex receive input from several sensory modalities. A major function of the cortex may therefore be the creation of poly-sensory models of the world, integrated with memory, in which the consequences of various courses of action can be simulated mentally.[4] We can be sure from our own experience that such models save a lot of trial and error in the real world, and they are probably particularly important in social species. Sociality implies a new complexity of behavior in which each individual interacts with other members of its group. This complexity is clearly evident in the functioning of a beehive, with divisions of labor and multiple forms of communication. In mammals it involves as well the need of individuals to assess the likely reactions of conspecifics to as simple a behavior as close approach, to say nothing of overt competition for food or mate. Put in these terms, the presence of the mammalian cortex suggests the possibility of some degree of conscious awareness, which by implication may therefore not be a uniquely human characteristic but one that has undergone a vigorous evolutionary development among social mammals.

Before venturing further into such uncharted waters, what can we say about the relationships between these several parts of the brain? The behaviors that are elicited by the hypothalamus and limbic system are not simple reflexes that occur with the predictability of a knee jerk. They are under complex control by other parts of the nervous system, and thus are strongly influenced by hormonal levels, the past history (memory) of the individual, and immediate sensory input. I am writing this on a warm summer morning on the terrace behind my kitchen. My two dogs have followed me outside (hoping for some excitement?), have sprawled in the sun on a stretch of a black asphalt path, and begun to doze. But within a minute of each other, the cortex of each is informed by her hypothalamus that the site is too hot. Their responses, however, are not the same. The older dog, prompted, I suspect, by memories of summers past, saunters inside to find the coolness of the hearthstone. A minute later the younger dog moves onto the lawn and into the shade of the big spruce tree where I have been contemplating a move myself. Still other solutions were possible, so why were these two selected? My point is that satisfying the simplest urgings of the hypothalamus generally requires that the central nervous system make decisions between alternatives.

Consider the richness of possible responses to hunger exhibited over time by any individual person. What to eat? How to eat it? Should I pick up the piece of chicken with my fingers? Who is watching? What will they think? There is an almost endless stream of factors that can shape a specific behavioral act. Only a fraction of these play any consequential role in determining the behavioral outcome, and on any occasion only an even smaller subset receives conscious attention. But those that do make us aware that there is frequently a struggle between the hypothalamus and limbic system's urging gratification and the neocortex's weighing other matters. There may be reasons for deferring a behavioral act, or for choosing between needs, or for responding in a particular way, all or any of which requires an assessment of experience, current information, and future consequences. In humans the process of assessment may or may not be a conscious, rational, deliberative act. In animals the process is traditionally not considered to involve conscious awareness. The history of this traditional view, as well as arguments for considering the possibility that animals are capable of some degree of thinking, have recently been discussed by Donald Griffin.[5] It is not necessary to adopt the position that animals think (or the alternative, that they do not think), however, to see the evolution of the neocortex as having provided an increased capacity to adjust behavior to a variety of environmental (including social) contingencies. Let us now put this argument back into an evolutionary context, where we can more easily see the possibility of natural selection at work.

The concept of behavioral scaling

The acorn woodpecker (*Melanerpes formicivorus*) is a bird familiar to residents of California, Arizona, and New Mexico. The name derives from its habit of plugging acorns into a matrix of nut-sized holes, creating a supply of food that it utilizes during the winter months. This unusual behavior is only one component of an interesting social structure. Typically the birds live and breed in communal groups of up to a dozen adults. They do not migrate, but occupy and defend a common territory. All members of the group are involved in feeding the young and in collecting acorns for their winter stores. What is most interesting, however, is that under certain conditions their social behavior can assume a very different form.[6] In certain parts of the range near the Huachuca Mountains of southeastern Arizona, where the acorn supply is not sufficiently reliable to support a wintering population, many of the birds do not form multimember groups, but associate in temporary pairs only for reproduction. Their storage of mast is casual and utilizes natural holes and crevices rather than specialized storage areas prepared for the purpose. And the birds migrate rather than winter over. As both kinds of behavior can be observed in contiguous areas, and as individual birds were observed to migrate one year and winter over with a group another year, the two forms of behavior represent alternative behavioral phenotypes that can be expressed by the same genotype. The critical environmental factor that appears to determine which pattern of behavior will be practiced is the relative abundance of acorns, which, in the Arizona study, was subject to annual fluctuation and could become insufficient to support the birds during the winter. In the words of the researchers who discovered this phenomenon, "the plasticity of the birds . . . may be an evolutionary adaptation to the marginal characteristics of the habitat in this area."

The flexible nature of the social structure of acorn woodpeckers is not an isolated biological curiosity, for examples of plastic social behavior can be found in species as diverse as honeybees and primates. In fact, the idea that evolution has produced behavioral capacities that can accommodate themselves to a range of environmental contingencies is commonly accepted by contemporary ethologists. The phenomenon has been formally termed "behavioral scaling" by E.O. Wilson.

> Behavioral scaling is variation in the magnitude or in the qualitative state of a behavior which is correlated with stages in the life cycle, population density, or certain parameters in the environment. It is a useful working hypothesis to suppose that in each case the scaling is adaptive, meaning that it is genetically programmed to provide the individual with the particular response more or less precisely appropriate to the situation at any moment in time. In other words, *the entire scale*, not the isolated points on it, is the genetically based trait that has been fixed by natural selection.[7] (Emphasis added.)

Although many of the examples that Wilson cites to illustrate this concept involve a quantitative scaling of aggression with either population density or the availability or quality of food resources, he very clearly had something larger in mind. Thus his definition of behavioral scaling includes changes in the "qualitative state" of behavior, of which the story of the acorn woodpeckers provides a paradigmatic example.

Note that Wilson's description of behavioral scaling is a statement about ultimate causation. It says nothing about the neurophysiological processes that are the proximate causes of behavior, except to infer that there are hormonal changes and sensory events that must transpire in order that alterations in behavior correlate with "stages in the life cycle, population density, or certain parameters in the environment." Social scientists are customarily interested only in matters of proximate cause, and their analyses are characteristically focused on those environmental events that correlate with behavioral changes. And some behavioral scientists of course focus on the underlying neurophysiological processes, which also contribute to explanations of proximate cause, but at another level. All three ways of studying such phenomena are valid. Moreover, the modes of explanation are complementary, and if the roles of proximate and ultimate cause are kept straight, the different approaches can interact synergistically to generate new and more powerful hypotheses.

Earlier in this chapter we saw that a central trend in the evolution of the vertebrate brain has been the development of mechanisms for allowing animals to adapt their behaviors to a wider variety of contingencies. In mammals, the cortical tissue that is involved in these processes is an evolutionary appendage, added over phylogenetically older neural tissue with somewhat more restricted functions and more directly related to the satisfaction of an animal's immediate needs. We are now in a position to appreciate that in an evolutionary context, drives or motivational states are examples of "proximate enabling mechanisms" by which the associated behaviors are made likely to occur. This evolutionary view of behavior also provides an interesting context in which to reflect on much that has been written about human nature by non-scientists, as for example the passage by the historian Barbara Tuchman that opened the chapter. Drives and emotions define only a part of the mammalian behavioral phenotype, but clearly not all of it. Comparative studies reveal a rich variety of ways in which the behavior of nonhuman species supports behavioral scaling—in which behavior is adjusted to the ecological contingencies at hand. But we have no logical reason for supposing that in the large and exquisitely complex brains we call our own the machinery for rational thought has been totally cut loose from evolutionary steering. There is really no reason for supposing that some of our most elaborate schemes are not responses to inner voices with their own agenda. Curiosity, ambition, pride, greed—these are unlikely to reflect the presence of hypothalamic control centers like those involved in hunger, thirst, and libido. Yet they are no less a part of

"*What is it, Roger? You seem lost in thought.*"

our being; they are components of our design; they contribute to the strategy of our genes.

All of which brings us to the making of decisions.

Animals as decision makers

Consider the general problems faced by an individual organism in making its way in the world. Focus further on a member of a resource-limited species, whose life is long (relatively), whose time must be divided between reproduction, including the prolonged task of rearing and protecting offspring, feeding, including the securing of resources, and avoiding the fate of becoming someone else's dinner. Life is an extended series of choices between a large number of alternative behaviors: whether to play, to fight, to feed, to feed on what, to explore, to court a mate, to mark a territory, to hide, to sing, to display, and so forth. Behavioral choices are not simply between competing drives—whether to feed or mate, fight or flee. Within any major category of behavior there is a myriad of secondary choices that must be made. A foraging animal may have several possible resources at its disposal, and it must decide how to allocate its time between them. What search pattern should a hummingbird adopt in gleaning nectar from a patch of trumpet flowers? How long should a hive of bees return to a given patch of clover, and what makes them shift their effort elsewhere? How should a pride of lions coordinate its hunting to cull a zebra from the herd?

Or contemplate in detail a very specific behavior, the "broken wing" response of killdeer to an approaching predator. Killdeer are a kind of plover, and they nest on open ground where they are susceptible to predation if their nests are discovered. The adults have a characteristic behavior to lure potential predators away from the area of the nest.[8] The bird first positions itself some distance from the nest and then attracts the attention of the predator by feigning injury, usually holding the wings as if broken and laboring across the field. If the dog, fox, or other animal fails to follow, the killdeer flies quite normally to a new location and tries the ruse again. If the predator follows, the killdeer continues to move away from the nest until either the other animal gets too close or the bird decides the nest is now safe. In either case it ceases to behave as though injured and resumes normal flight. Although dogs and humans can elicit this behavior, not all animals do. Large grazing animals are more of a danger to the nests by walking on them, and killdeer have an alternative behavior for dealing with this threat. They remain close to the nest and become as conspicuous as possible, thereby causing the cow or sheep to step aside.

This process of defending the nest requires the central nervous system of the bird to assess a stream of sensory information and generate a variety of behavioral responses. In short, the killdeer must make a number of decisions. First it has to decide what kind of a threat the approaching animal poses to the nest—will it tramp on the eggs or young, or is it more likely to eat them? Assuming the latter category, how far from the nest should the parent killdeer go before it reveals its presence? How close should it let the predator approach before taking flight? Alternatively, is the ruse working; is the fox following? Has it been lured far enough from the nest; is the danger over? In some sense, each of these questions must be answered at least once during the enactment of this behavior.

The reader may be bemused by the suggestion that the bird is making a series of decisions. This is because so much of the fine-tuning of human behavior in similar circumstances appears to involve conscious choice. Whether the bird is aware of its actions is, from considerations of ultimate cause, no more important than the knowledge of which neurotransmitters are involved. The bird acts as if it were making decisions by similar criteria that you or I would use if faced with the same problem and the same means of dealing with it. It is the assumption that only humans possess conscious awareness that has created the expectation that animal behavior must be a simple, one-dimensional process that lacks plasticity and a capacity to change adaptively as circumstances alter. Yet, by some process, conscious or otherwise, the central nervous system of the killdeer is making decisions. This capacity for selecting the apparently appropriate response on a landscape of various possibilities is part of the proximate enabling mechanisms contributed by the evolution of the brain. It is part of the strategy of being a killdeer.

What an animal chooses to do at any moment depends on an extensive array of factors: internal physiological condition, such as nutritional and reproductive state, and some assessment of advantages and disadvantages. Moreover, there may be very specific advantages and disadvantages associated with every choice open to it. Worse, the ultimate advantages and disadvantages may not be apparent either to the animal or an observer, so many decisions will be fraught with ambiguity or conflict. Yet there can be no question that animals do make such choices continually.

But what is meant by advantage and disadvantage? And how are they measured? This is a complex question in which it is necessary to recognize both proximate and ultimate aspects. The evolutionary history of the species endows each individual animal with a range of behavioral possibilities that, in the aggregate, serve to enhance its survival and reproductive success. This is one manifestation of what we earlier referred to as the organism's design for fitness. A particular behavioral choice, however, could increase, decrease, or have no effect on survival or reproductive success. Moreover, survival and reproductive success are obviously not the same thing; early in life, survival of self may be a prerequisite for reproductive success, whereas later in life sacrifice of self might contribute to inclusive fitness. Throughout their lives individuals will exhibit patterns of choice that enhance reproductive success and survival, but individual choices may turn out to be mistakes. If the choice is fatal, the game is very likely over. If the mistake is of a less serious nature, the experience may be stored in memory and influence subsequent choices. Therefore natural selection can function either to enhance or eliminate certain behaviors with a large amount of genetic programming, or it can function to increase learning capacity, or as we have seen, it can function to increase the capacity to learn very specific sorts of associations. With respect to the outcome of natural selection, it makes little difference whether the mechanism by which learning becomes directed involves changes in the sensory nervous system or in the neural processes by which associations are formed. In principle, both avenues are open.

Evolutionary theory predicts that there should be some rules that govern the processes by which animals make decisions, and one of the major thrusts of current behavioral ecology is the testing of hypotheses that specific behaviors are chosen to optimize particular parameters, as, for example, caloric input and energetic costs in foraging. This mode of analysis is addressed to ultimate causes, in terms of adaptation and evolutionary history, and can provide insight into those factors that have been most important in shaping the evolution of a species. It may not address, however, the related question of the proximate causes of behavioral choices. In this arena animal behaviorists who pose evolutionary questions are just beginning to interact with their counterparts who come from psychology.[9] What are the sensory cues and the cognitive processes that underpin decision making? How are the memories of past experience weighed in the process of choice?

This application of optimization theory has been criticized on the grounds that it assumes that every behavior is adaptive, so every case that is examined becomes a self-fulfilling prophesy.[10] Proponents of the approach,[11] however, have argued effectively that to suggest that behavior optimizes a particular parameter (e.g., caloric input per energetic cost in foraging) is to state a very specific and falsifiable hypothesis about the relationship between a limited number of defined variables. Although correct, there is a limitation to this approach. The validation of a specific hypothesis is likely to depend on environmental conditions or other factors that may remain unrecognized by the experimenter. The generality of any conclusion must therefore always be further tested in order to define the relevant boundary conditions. In short, what (if anything) is optimized will likely depend on circumstance,[12] all of which is a roundabout way of saying once again that the contribution of a behavior to fitness is dependent on the context in which the behavior occurs.

Decision makers as animals

The very general description of the behavior of a resource-limited species that I have given in the preceding paragraphs can be applied to *Homo sapiens* but with some additional provisos. First, the evolution of language has enriched by orders of magnitude the complexity of social organization over both space and time. Thus experience relevant to any decision may come as information from previous generations or from contemporaries as well as from the personal history of the individual. Similarly, the consequences of any decision may spread in manifold ways through the social structures of which the individual is a part. All of this complicates enormously the decision-making process, enriching it with much more contradiction, ambiguity, and uncertainty than exist for any other species. Furthermore, humans are not only very good at recognizing other individuals by their faces, but this ability is supplemented by faculties for assessing emotions, intentions, and their degrees of sincerity. Deceit in matters large and small is very much a part of human relations. At the same time, however, those same threads of ultimate causation that can be traced through the fabric of comparative animal behavior are woven into the natural history of *Homo sapiens*. In the next chapter we shall illustrate this point with reference to the literature of anthropology, but at this juncture let us consider the phenomenon of aggression.

One often hears the argument that humans cannot be innately aggressive because they also exhibit such a capacity for cooperation. (One never hears the converse—that people cannot possibly be innately cooperative because they are so aggressive—but it would make about as much sense.) Stephen J. Gould, in a critical review of E.O. Wilson's *On Human Nature*,[13] has illuminated the essence of this seeming paradox while at the same time finding his own formula for obfuscation.

> The critics of sociobiology do not seek to deny the importance of biological factors in human nature. I believe, however, that Wilson has made a fundamental error in identifying the wrong level for biological input. He looks to specific bits of behavior and their genetic advantages, and invokes natural selection for each item. He tries to explain each deed rather than the underlying ground that organizes an act as one mode of behavior among many. . . . Thus Wilson asks: "Are human beings innately aggressive? This is a favorite question of college seminars and cocktail party conversations, and one that raises emotions in political ideologues of all stripes. The answer to it is yes." As evidence, Wilson cites the prevalence of warfare throughout history. "The most peaceable tribes of today," he writes, "were often the ravagers of yesteryear and will probably again produce soldiers and murderers in the future." But if some peoples are peaceful now, then aggression itself is not encoded, only the *potential* for it. If "innate" means only possible, or even likely under common circumstances, then it cannot bear Wilson's claim that natural selection works to choose the best alternative. We should seek a biological basis in the generating rules of human behavior, not in specific actions.[14]

In this passage Gould correctly sees aggression as one of a number of possible behavioral responses of humans—as indeed it is for many animals. Where his critique seems to me to go seriously awry, however, is in the shallow view of behavioral evolution he attributes to Wilson. There is in fact a great deal of similarity between Wilson's description of behavioral scaling that was quoted earlier and Gould's elegantly phrased suggestion that "we should seek a biological basis in the generating rules of human behavior, not in specific actions."

Human beings are as "innately aggressive" as other vertebrates. Just as people feel hungry when their bodies need food, they feel angry and aggressive (and/or fearful) when they perceive a threat to pride, mate, resources, offspring, or some other important aspect of existence. There are measurable responses of the nervous and endocrine systems that accompany these feelings, and the responses have their homologous counterparts in other vertebrates. Aggressive motivations are therefore one of the evolutionarily conservative devices with which animals are equipped to ensure their survival and promote their reproduction. To deny their existence in animals is to close one's eyes to an enormous amount of comparative physiology and behavior, and to deny their existence in humans flies in the face of common experience as well as common sense.

The dialectics of human aggression (and other aspects of sociobiology) can be supremely inane. Some very sophisticated individuals on the political left have been concerned that an evolutionary perspective on human behavior is dangerous and can lead to a confounding of what is "natural" with what is "right" for society. "Survival of the fittest" is indeed not a formula for social harmony or social justice. Others, less sophisticated, while failing to grasp the scientific or even the metaphysical issues, have rallied to the cause and dismissed evolutionary reasoning as the work of latter-day Calvinists, or worse. All of this is a great pity, because it contributes not a whit to understanding either the evolution of the human psyche or current social ills.

Among social scientists, aggression is sometimes characterized as either puerile or primitive. Although masquerading as statements of cause, these are little more than value judgments. With very few exceptions, the violent behavior of individuals is unwelcome within any social group and is proscribed by formal rules of conduct. To attribute aggression to the socially or biologically undeveloped is simply to reinforce a desired norm for the society. As an explanation of violence, it fails. Most of the world's serious mayhem is performed at the hands of adults, and the more developed the culture, the more efficient the death and destruction.[15]

The topic of aggression, however, gives us a wedge to pry into a much larger set of issues. The human behavioral design that we have been trying to characterize is one that must serve the purposes of individuals each of whom has a unique set of reproductive interests. Because each is unique, conflicts between individuals are inevitable. In a very fundamental sense, people are selfish in that they act in what they perceive to be their own best interests. Because we are long-lived, reproduce repeatedly, and our young require much care, our reproductive interests are fostered by our capacity to garner and control resources. So it is no surprise that resources figure prominently in our perception of our own best interests.

On the other hand, because we are social creatures living in groups, no individual's interests can be realized without some cooperation from other members of the group. Herein lies the key to the apparent paradox of altruism. If groups were invariably small enough that all members were related, seemingly altruistic acts might simply improve the altruist's inclusive fitness. It is much more likely, however, that seeming altruism has a more subtle psychic base. In a society of individuals each of whose success depends on cooperation, an act extended to another at cost to self is likely to be reciprocated in some way at a later time. Why is it likely to be reciprocated? If something is done for you, it pays you to repay, lest you fail to receive help on another occasion. Or perhaps to be more accurate, it pays you to convince others that you are reliable and are likely to repay, particularly if you can accomplish these ends at minimum cost to yourself. Your interests, both reproductive and somatic, are best served if you have the goodwill of your peers, for you are more likely to be assisted in the future. Your interests are also well served if you are skillful in detecting when others are deceiving you.

A number of seemingly contradictory aspects of the human character follow quite naturally from this reasoning: among them are greed, acquisitiveness, concern for one's image, pride, and consideration for others. The conflicts between these impulses are generated by immediate circumstances, and the conflicts can generate real problems. Nevertheless, these inclinations and propensities are all useful instruments in the behavioral tool kit that has gotten us to this point in our history.

It is not necessary to recognize consciously what one's genetic interests really are, or that in a very fundamental way the genes are the driving engine. As Richard Alexander[16] has pointed out, we have only known about genes for about 100 years, and it is neither possible for us to have understood ourselves before that discovery nor necessary that evolution should have provided our consciousness with its secrets.

The unconscious was elevated to a position of prominence in discussions of the metaphysical by Freud, and it is probably time to take a more evolutionary view of his ideas. (As an example, it has been suggested that the "Oedipal conflict" is not overt sexual competition between father and son for the mother but is a more general manifestation of parent-offspring conflict over the degree of maternal investment.[17]) What may be of most lasting importance in Freud's contributions, however, is the recognition that the reasons for our behavior need not necessarily be obvious to us. Evolution has produced a psyche that is adept in building coalitions with others, that is curious about the world and is therefore effective in exploiting what resources are available, and that is propelled in these ventures by stimuli that create an inner state of satisfaction or pleasure. It may think it knows what it is doing when it attends to self, but it is frequently responding to silent voices.

Back to aggression. The question of when, under what circumstances, and how aggressive behavior is expressed is more complicated than the matter of its existence. Aggression usually arises in response to a perceived threat. In terms of aggressive encounters between individuals, not only is the perception of the threat important, but so is the assessment of the consequences of various actions. Both are in turn strongly influenced by the training that is imposed by the social system of which the individual is a part, as well as by immediate physiological conditions (such as state of rest or hunger). But none of these caveats means that aggressive urges do not exist under more or less predictable conditions and that aggressive behavior does not frequently follow. Once again, we find that the cortex intrudes, sometimes to censor the hypothalamus, sometimes to conspire with it. The human brain can turn simple urges into extensive schemes accompanied by elaborate justifications to self and others. Aggressive behavior may be subtle or overt, personal or pervasive.

The case is strong that the human brain has been molded through the interplay of selfish and cooperative behavior within groups and supplemented by intense competition between groups. Human history is largely the record of struggles for control of resources by one group at the expense of others. As I write these words I am prompted to list the places in the world where political groups of people are now at one another's throats. With no pretense of being comprehensive, this morning's paper brings news of Catholic against Protestant in Northern Ireland, Arab against Jew in the occupied territories of Israel, Arab against Arab in Kuwait, intertribal conflict in Liberia, civil wars in Cambodia, Afghanistan, and Ethiopia, Tamil against Buddhist (and Muslim) in Sri Lanka, Hindu

against Sikh in India, Arab against Kurd in Iraq, Mohawk Indian
against whites in Quebec, landowner against peasant in El Sal-
vador; and ready to detonate: Hindu against Muslim in Kashmir,
black against white in South Africa, and countless ethnic rivalries
throughout Eastern Europe, the Soviet Union, and sub-Saharan
Africa. All this, while part of the world prides itself for having
been at peace for nearly fifty years.

I submit that it is hard to name a major political altercation,
past or present, that does not come down to a conflict over
resources, whether they be land, water, minerals, or, where the
combatants occupy the same turf, access to jobs or a more
appropriate share of the group's wealth. Frequently the conflict
may appear to be about something else, such a religion or race,
but these are usually just the tokens by which one group of combatants distinguishes itself from its rival in its quest for relative
power, control, and ultimate reproductive success. As a symbol
of a way of life, religion can have a powerful hold on people,
as in the English civil strife of the seventeenth century or in Iran
or northern Ireland today.

In all these conflicts the players behave as though life were a
zero-sum game and one group's success must come at the expense
of another's. The groups see themselves as sharing common interests and the members are therefore bound by some form of loyalty,
often reinforced or justified by political and religious leaders, and
cast in terms that suggest the rival is unworthy. As A. Bartlett
Giamatti observed to graduating Yale seniors several years ago (although I suspect he did not intend to be heard as an evolutionary
psychologist), " . . . an ideology can encourage people to murder
as easily as it can encourage them to nobility."

That much of human evolution seems to have been driven by
people cooperating and competing within groups and competing
between groups for access to resources has a great deal to do with
how we see the world and how we relate to each other. Richard
Alexander has pointed out that the very existence of ethics and
morality arises from the fact that individuals are genetically unique, therefore pursue different goals, and of necessity have conflicts of interest. "Without conflicts of interest . . . the very
concepts of ethical and unethical, moral and immoral, and right
and wrong would not exist." Contrary to the conventional view,
morality does not require self-sacrifice; "moral systems [are] sys-

"We do pretty well when you stop to think that people are basically good."

tems of indirect reciprocity"[18] that function to orchestrate individual behavior in such a way as to dampen intragroup conflicts while inflicting a minimum cost if not a net gain (in evolutionary currency) to individual members of the group. The Ten Commandments address such issues quite directly, with admonitions about killing (which in the bloody context of the Old Testament we can safely assume refers to members of the same group, identified by worship of the "right" God), adultery (very disruptive), stealing (again the context indicates that the proscription applies to members of the group), bearing false witness against one's neighbor (by definition, intragroup), and coveting one's neighbor's possessions, including his wife.

The concept of morality's producing the greatest good for the greatest number is consistent with evolutionary principles only when the interests of the individuals are very similar. This has probably frequently been the case with small homogeneous groups in competition with other groups; it is much less obviously so when the groups are large and heterogeneous. When groups are ruled by a small elite, the rules and moral systems tend to work to the interests of the rulers. Earlier we saw the direct correlation between power and reproductive success in a sample of the most despotic societies known. A trend in larger groups has been to defuse the conflict inherent in polygynous systems by the legal encouragement of monogamy and to leave to individuals the right to arrange their own reproductive lives. In contemporary American society there is a bitter irony in the spectacle of those who wish to perpetuate the paramount importance of the family while also working to impose political control over the reproductive choices of women.

In summary, natural selection works on individual organisms, not individual behaviors, and whether a behavior enhances fitness depends on circumstances. We are a social animal whose intelligence allows us to exploit a variety of habitats, which is another way of saying that we have evolved a pattern of behavior that is sufficiently flexible to deal with a spectrum of contingencies. The nature of that spectrum, how individuals respond, and the conditions which, statistically, generate particular classes of response are proper subjects for inquiry. If particular social conditions can be counted on to produce behavior that consensus deems undesirable, a reasonable strategy is to change the conditions. This is the complete antithesis of genetic determinism, and as Richard Alexander has suggested, "evolution is surely most deterministic for those still unaware of it."[19]

8
Culture, Anthropology, and Evolution

> The great flood which had covered the earth for so long had at
> last receded and even the thin strip of sand stretching north from
> Naikun . . . lay dry. The Raven had flown there to gorge himself
> upon the delicacies left by the falling water, and so for a change
> wasn't hungry. But his other appetites, lust, curiosity, the
> unquenchable desire to interfere and change things, to play tricks
> on the world and its creatures, these remained unsatisfied.
>
> From the Haida legend of the
> Raven and the first humans.

Coevolution of biology and culture: the "leash effect"

Although culture has traditionally been viewed by many anthropologists as a uniquely human attribute, recent work in animal behavior has demonstrated a number of examples of behaviors that
appear to be learned by imitation and to be localized to separate
populations. A famous example is the habit of separating grains of
wheat from sand that was invented by Imo, the young female
Japanese macaque that passed on the trick to other members of the
troop. The monkeys were provided with wheat grains on a beach,
and Imo found that if she threw a handful of sandy wheat into the
water, the sand would sink and the edible seed could be skimmed
from the surface. Imo's age peers were the first to learn the tech-
nique, but in time mothers taught their offspring.[1] Other examples
can be cited in a large number of species.[2,3] As a consequence of
these observations, there have been efforts to sharpen the concept
of culture so that it can be studied as a general biological
phenomenon, and this has led to much new discussion of the
relationship between organic and cultural evolution.[4]

There are at least as many definitions of culture as there are
textbooks of anthropology. Paul Mundinger[5] has reviewed the his-

tory of the term and attempted to formulate a definition that is consistent with both mainstream use in anthropology and the idea that a number of nonhuman species exhibit at least a protoculture. The sense in which I shall use the word is similar to his: Culture consists of those patterns of behavior (and their neural codes without which the overt behavior could not exist, as well as the material products of the behavior which may be the only information we have about cultures of previous times) that are acquired by observational and imitative learning from other members of the species and which are replicated generation after generation. Because the concept of culture is intended to deal with the variation in behavior between populations of conspecifics, and because in humans there is much cultural diffusion between populations, there is a strong presumption that, in general, different cultures are not based on differences in genotype. Even when members of different human populations look different and where there are demonstrable differences in gene frequencies underlying physical characteristics, the presumption remains that cultural differences reflect alternative phenotypic expressions of a common genetic heritage.

Although this traditional presumption is probably nearly always correct, the possibility of genetic differences contributing to cultural variations cannot be casually dismissed. For example, the low use of dairy products by some populations of people is very likely related to the known difficulty of the adults of those groups to digest milk. In principle, therefore, it is not absurd to consider, for example, the possibility that populations of people from different geographic regions may exhibit statistically different patterns in the biochemistry of the neuro-active molecules of the brain, with consequent differences in personality or other behavioral characteristics. Even with this caveat in mind, however, as a general explanatory proposition to account for cultural differences, genotypic differences are not the answer. Moreover, I know of no biologist—socio- or otherwise—who believes that they are.

A brief aside: Mundinger draws a distinction between tradition and culture that holds a lesson. In his classification, traditions differ from cultures in that the latter require an overt model of the behavior generated by conspecifics that can be imitated by novices. As an example of an animal tradition that is not a culture, he cites the use of specific olfactory cues by spawning salmon in locating the streams of their birth. This concept of tradition runs into difficulty with the assertion that

> ... two different criteria can be applied to identify as traditions, animal behaviors described in the literature: (1) evidence that a particular behavior is passed on from generation to generation by learning and/*or* (2) evidence that different forms of a modifiable behavior characterize different populations of a species.[6] (Emphasis added.)

The first criterion focuses on learning as though it were always readily dissected from the rest of the ontogenetic process, and standing alone it might be construed to include even imprinting. The second criterion assumes that because populations differ, the cause must be differential learning. Different populations might express different behaviors (a) because of different learned traditions, or (b) because they are genetically different, or (c) because different environmental conditions prevail in the two geographical areas, eliciting different behavioral responses from individuals from equivalent gene pools and without the involvement of any imitative learning whatsoever. The acorn woodpeckers discussed previously may offer an example of the third possibility.

Table 1 compares some of the features of organic and cultural evolution in a manner that illuminates their differences as well as their interaction. The essential differences are as follows: In organic evolution the fundamental unit of replication is the gene, but genes are packaged in organisms. The process of natural selection—the differential reproduction of organisms possessing different assemblages of genes—along with the other contributing causal mechanisms of evolutionary change, result in a differential transmission of alternative alleles to succeeding generations. We can thus distinguish between the replicators (genes) and the things that contain ensembles of replicators but which actually get selected (phenotypes). In cultural evolution, on the other hand, the replicating elements are behaviors and their neural substrates, and these are the same entities that are selected in cultural transmission.

Obviously cultural evolution can proceed with no differential transmission of genes. "Psychological selection"—the differential transmission of alternative behaviors—can be rapidly propelled by selective teaching of the young by their elders. Moreover, teaching of alternative behaviors can proceed in ways that are indifferent to

TABLE 1

Parallels Between Organic and Cultural Evolution
(Modified from Mundinger[7])

DEFINITIONS AND COMPONENTS OF MICROEVOLUTION	ORGANIC EVOLUTION	CULTURAL EVOLUTION
Evolution	Changes in the relative frequencies of genes over time	Changes in relative frequencies of culturally determined behaviors (and their neural codes) over time
Replicator	Genes	The culturally determined behaviors (and their neural codes)
Mechanism of transmission	The reproductive process	Imitation, broadly interpreted
Source of variation	Mutation	Copy-error; invention
Causal mechanism of evolutionary change	Natural selection	"Psychological" selection
	Gene flow	Diffusion of culture
	Genetic drift	Cultural drift
	Mutation pressure	Species-specific behavioral predispositions (psychic unity)
Agents of selection	Differential births, deaths, and matings	Differential learning predispositions and differential tutoring
Proximate unit of selection	Phenotype	Specific phenotypic characters (behaviors)
Ultimate unit of selection	Gene	Specific phenotypic characters (behaviors)

natural selection. Whether a people speak Chinese, German, or Arabic and eat with chopsticks, forks, or fingers seems unlikely to have much to do with fitness. These are all viable alternatives.

This having been said, it is perhaps more instructive to consider the points of intersection of the two evolutionary processes. Culture is one aspect of the behavioral phenotype; therefore, at some fundamental level it is tied to, and dependent on, genotype. This relationship may account, at least in part, for the reappearance of certain cultural forms in different geographical areas, sometimes as responses to common ecological problems, or sometimes reflecting predispositions to learning. Such a general idea lies behind the concept of "human nature" as well as efforts to paint ethograms for the human species, and is consistent with certain elements of "structuralist" theories in anthropology.[8] It is clearly more useful in addressing what is common to the human condition than it is in accounting for differences between cultures. But what about differences?

One of the more promising themes in anthropological theory involves efforts to view organic and cultural change in a coevolutionary framework. Many cultural choices represent options that, like adaptations, enhance reproductive success and survival. Indeed, many anthropologists see culture as the means by which people come to terms with their environment—a pattern of behaviors that enable them to garner resources, insure their survival in the face of unpredictable natural cycles and competing neighbors, establish order within their societies, win the approval of conspecifics, and attempt to generate some degree of peace in the psyche. What is the engine that drives this process? On the one hand there appear to be some broad instructions written in evolutionary time, such as looking after the well-being of offspring or being concerned with the reproductive activity of mates. The particular ways in which these interests are expressed may vary greatly, and ecological or historical factors may be the proximate determining agents in creating differences between cultures. In these instances evolution has contributed only a very general guiding hand. To supplement whatever behavioral tendencies characterize humankind, we are also endowed with the capacity to anticipate and to benefit from the experience of self or others. In fact, human language is such a highly developed system of communication that we can draw on the experience of individuals far removed in geography or time. Behavioral choices may thus appear to be adaptive in the sense of enhancing fitness, yet not involve adaptation per se. In other words, alternative cultural solutions to similar ecological problems may be invoked with no alteration in the relative frequencies of genes.

But the interaction between cultural and organic evolution can be cryptic. We have seen how natural selection can focus what is learned, thus shaping one of the components of psychological selection and helping to mark the boundaries of the behavioral repertoire. But conversely, what is learned is also a function of environmental demands, and if the demands change, so does the repertory of behaviors. What is often imperfectly appreciated is that because patterns of behavior can be important determinants of such issues as mortality and the selection of mates, behavior—even culturally acquired behavior—can, in principle, have an influence on the gene pool. This feedback can either reinforce existing learning predispositions in the population, or in a changing environment it can reinforce shifts in gene frequencies brought about by natural selection. Thus cultural and organic evolution meet in the arena of development—where learning plays a final part—and where once again we encounter the subtle interplay of genetic and epigenetic events. It is in this sense that we can speak of the coevolution of culture and organic change.

Charles Lumsden and Edward O. Wilson[9] have approached this problem from a somewhat different perspective. Suppose, for the sake of argument, that natural selection had created a species that was truly a tabula rasa organism whose choices of cultural alternatives were totally unbiased by any genetic constraints. Under these conditions, any cultural choice could occur without regard to its effect on genetic fitness, but the system would be unstable in evolutionary time. It would be unstable because any genetic change that tended to bias the individuals possessing it toward cultural choices that enhanced genetic fitness would be bound to increase in frequency by natural selection. This is the sense in which the genes "keep culture on a leash."

This hypothetical initial tabula rasa state might be perturbed by either of two processes. We have just seen that new genotypes that affect the probability of cultural choices could alter fitness and spread through the population. But it is also true in principle that cultural choices, even when not the result of any genetic steering, can alter the fitness of different genotypes. An interaction of cultural and organic evolution therefore seems highly probable, if not inevitable.

Sociobiology and cultural materialism

Among the various theories invoked by anthropologists to provide a guiding framework for their discipline, cultural materialism, as expounded by Marvin Harris,[10] has some appealing features for a biologist. At the same time, however, it provides another example of my contention that evolutionary biologists have not managed to communicate their message with sufficient clarity to colleagues in related disciplines. I have included the following brief section on cultural materialism in response to Harris's "hope that the advocates of [alternative] strategies will be moved by the possible biases of my interpretations and that they will seek to clarify, if not to change, their positions."

What is the common ground that is shared by evolutionary biology and this school of anthropology? In describing cultural materialism, Harris goes to some length to compare it with other systems of explanation in cultural anthropology. In contrast to most of the alternatives, he asserts that sociobiology and cultural materialism share a "forthright commitment to the general epistemological principles of science, [and] in this respect [they are] natural allies." His approval, however, stops there.

Cultural materialism seeks to identify the causes of cultural choice and diversity first in the infrastructure of human society—the modes of production and of reproduction. Although there are Marxist roots to this system of explanation in its emphasis on the modes of production, Harris justifies it on grounds that need only relatively minor editing to reflect much biological insight:

> Like all bioforms, human beings must expend energy to obtain energy (and other life-sustaining products). And like all bioforms, our ability to produce children is greater than our ability to obtain energy for them. The strategic priority of the infrastructure rests upon the fact that human beings can never change these laws. We can only seek to strike a balance between reproduction and the production and consumption of energy."[11]

This is indeed the fate of a relatively large, long-lived, social, serially reproducing, resource-limited animal. Consequently, it is easy to find merit in a theory of cultural diversity that places primary emphasis on the relations between human populations and the ecological systems in which they exist. For example, cultural materialism postulates that the sizes of bands of hunter-gatherer peoples, their need to adjust local population density seasonally, the nuclear family as a unit of band structure, and features of social organization such as exchange of goods and marriage between bands all flow from the nature and distribution of the resources on which the people live. Similarly, a substantial increase in group size requires an intensified mode of production, such as a shift to agriculture. The opportunity for agriculture based on irrigation leads to large imperial systems more readily than in geographical regions where growing crops depends more immediately on rainfall. And the different pace of evolution of cultures in the Old and New Worlds is postulated to be related to the Pleistocene extinctions of domesticatable ungulates in the western hemisphere. These are but brief allusions to a detailed and thoughtful account of a cultural history whose texture cannot be adequately conveyed in a single paragraph.

Cultural materialism also recognizes, in a vague way, that there are "bio-psychological principles" that drive human behavior and that are shared by other primates. Harris's minimal list is kept as short as he can make it:

1. People need to eat and will generally opt for diets that offer more rather than fewer calories and proteins and other nutrients.

2. People cannot be totally inactive, but when confronted with a given task, they prefer to carry it out by expending less rather than more energy.

3. People are highly sexed and generally find reinforcing pleasure from sexual intercourse—more often from heterosexual intercourse.

4. People need love and affection in order to feel secure and happy, and other things being equal, they will act to increase the love and affection which others give them.[12]

One can argue about the composition of this list; its existence, however, is tacit recognition that there are some evolutionarily conservative elements to primate behavior. The door is open; evolutionary history is acknowledged.

Harris has been critical of sociobiology, viewing it, like other anthropological theories, as a strictly competitive system of hypotheses. As I indicated above, he enters this discussion because I believe his aim is wide of the mark, and I hope that understanding can be improved by considering his critique in more detail.

His first criticism would be devastating if it had any relevance. Like the remarks of another anthropologist that were quoted at the start of Chapter 4, however, it conveys a view of biology that is a hollow caricature:

> The weakness of human sociobiology and all other varieties of biological reductionism arises initially from the fact that genotypes never account for all the variations in behavioral phenotypes. Even in extremely simple organisms, adult behavior repertories vary in conformity with each individual's learning history.[13]

This would be a weakness only if biologists attributed such authority to genes. They do not, of course, and curiously, Harris actually seems to be aware, for five pages later he writes:

> Popular representations of sociobiology have created a false impression of how sociobiologists relate human social behavior to its genetic substrate. Sociobiologists do not deny that most human social responses are socially learned and therefore not directly under genetic control. Wilson . . . has made this point without equivocation: "The evidence is strong that almost but probably not quite all differences among cultures are based on learning and socialization rather than on genes." Richard Alexander . . . has made the same pronouncement: "I hypothesize that the vast bulk of cultural variations among peoples alive today will eventually be shown to have virtually nothing to do with their genetic differences." Thus few if any sociobiologists are interested in linking variations in human social behavior to the variable frequencies with which genes occur in different human populations. . . . [14]

But let us return for a moment to the passage on the "weakness" of sociobiology. Harris here pays lip service to the fundamental inseparability of the learning process and the evolutionary background on which it occurs, yet he never fully incorporates the significance of this truth into his analyses. Thus genetic and environmental causes are viewed either as alternatives, or the idea that a particular behavior has been molded by evolution is seen as a "redundant and gratuitous" additional hypothesis. These mistakes follow, respectively, from an insufficient appreciation for the complexity of interaction of genetic and epigenetic events as proximate causes of behavior, and a failure to distinguish between the concepts of proximate and ultimate causality. Harris is right to view the evolution of culture in an ecological context, but he is unwise to separate cultural anthropology so remorselessly from evolutionary biology. This is not the path to understanding, for a scientific explanation, however parsimonious, loses all elegance—to say nothing of importance—when it ignores a significant aspect of nature. One can shave too closely with Occam's razor.

Harris's second and third criticisms illustrate these points. The second has to do with the content of the human "ethogram"—with the nature of human nature. While acknowledging that humans have certain species-specific behavioral propensities (which were listed above), Harris would like to keep the list as short and as general as possible. On this matter his disagreement with biology is therefore a quantitative rather than a qualitative issue. His position is tied to that set of traditions in the study of human behavior that assumes there has been no evolutionary channeling of cognition and learning and rejects the hypothesis that certain behaviors are more probable than others because of the evolutionary history of the species. As pointed out above, this position is theoretically unsound because it cuts off the study of cultural evolution from the rest of biology and precludes examination of the problem in every dimension. And it is most easily maintained when the concepts of proximate and ultimate cause become confused, which brings us to Harris's criticism of behavioral scaling.

We have seen earlier that one of the general features of animal behavior is that what animals do frequently depends on the circumstances in which they find themselves. The circumstances, in turn, are made up in various proportions of immediate environmental stimuli, past experience, and evolutionary heritage. Even where there is a large component of genetic programming, the behavioral response is frequently tuned to the specifics of the moment. The concept of behavioral scaling has not been grafted onto the study of human behavior to account in genetic terms for behavioral plasticity. The idea arises in animal behavior from considerations of ultimate cause, but it embodies no specific assumptions about the mechanisms of proximate cause.

The confusion of proximate and ultimate cause is vividly apparent in the following passage:

> True, sociobiological models based on reproductive success and inclusive fitness can yield predictions about sociocultural differences that enjoy a degree of empirical validity. . . . But the reason for this predictability is that most of the factors which might promote reproductive success do so through the intermediation of *biopsychological benefits* that enhance the economy, political, and sexual power and well-being of individuals and groups of individuals. . . . Thus sociobiology contributes to the obfuscation of the nature of human social life by its commitment to the exploration of least probable causal relationships at the expense of the most probable. [15] (Emphasis added.)

Without an evolutionary context, "biopsychological benefits" have the same ad hoc character that Harris deplores in the structuralist theories of Levi-Strauss. In fact, they are among the proximate enabling mechanisms through which the genes speak. As Harris implies, the relation between the satisfaction of psychological urges and the "need" or "purpose" of the behavior that produces gratification is frequently very obvious. We take for granted that people must eat and reproduce. But the concept of psychological benefits as proximate enabling mechanisms will be clearer if we consider an example where the "purpose" of the behavior may not be so obvious as our intuition leads us to believe. Consider sleep. When we are deprived of sleep, we feel drowsy—we experience a psychological need for sleep. Moreover, as drowsiness impairs our ability to perform tasks with accustomed facility, and as feeling tired is reversed by sleep, we conclude that the purpose of sleep is to provide a needed period of recovery, a recharging of the batteries. But it is not obvious that this explanation for why we sleep is correct.[16] First, people vary greatly in the amount of sleep they require, and some individuals thrive on as little as one hour in twenty-four. Furthermore, if we look at other mammals we find that there is enormous variation: bats may sleep as much as twenty-two hours a day, and many of the large grazing mammals less than three. Is it possible that vertebrates differ so widely in their need for periods of biochemical and physiological recovery? Or is the "purpose" of sleep to synchronize periods of activity with the solar day in a manner that varies with the feeding behavior, susceptibility to predation, and other aspects of the ecology of each species? Is hibernation but an extreme form of this behavior? In this view, "it is the sleep control mechanism that makes us feel tired,"[17] and we have misinterpreted the meaning of sleep. Feeling drowsy is a proximate enabling mechanism to produce a specific behavior, regardless what purpose we believe the behavior serves.

Let's consider an example with more anthropological interest. It is a delusion to suppose that the wide variation in marriage practices observed throughout the world is evidence that there has been no influence of evolution on human mating behavior. Not all conceivable forms of mating occur, and those that are observed do not occur with equal frequency. When Harris writes that " . . . people can be socialized into and out of promiscuity, polygyny, polyandry, and monogamy with conspicuous ease, once the appropriate infrastructural conditions are met," he is correctly pointing out that human sexuality has several phenotypic expressions that can be elicited under various conditions. The argument, which is intended to dispose of the notion of a species-specific human mating behavior, does not refute the hypothesis that our species has a significantly polygamous character, uneasily fused together in evolutionary time with powerful, if conflicting, tendencies for pair-bonding. It is probably no accident that monogamous societies have to resort to considerable "socialization" to reinforce their marriage codes, or that rampant promiscuity is not a stable cultural alternative. The argument also does not address the nature of the infrastructural conditions and how they lead to various mating practices, but in general I suspect that the proximate causes that would be advanced by a cultural materialist would in fact be in harmony with a somewhat broader perspective that sees cultural and biological change frequently interacting in a coevolutionary process. Let us illustrate with a specific example.

The case of Tibetan fraternal polyandry

Polyandry, the marriage of more than one man to an individual woman, is a relatively rare practice in human populations. In fact, as we saw in Chapter 4, it is an uncommon mating system among all animals. When it occurs in human populations it usually takes the form of fraternal polyandry, in which the men involved are brothers. Fraternal polyandry in a Tibetan population in Nepal has been the object of close study by anthropologists who have used their analysis to reject what they believe to be a sociobiological explanation for this marriage custom.[18]

What are the facts of the case? Polyandry occurs in remote, isolated villages in which land is scarce and valuable. Far from being a universal practice even among these people, polyandrous marriages are made only by land-owning brothers, and then

> to preserve and increase the productive resources (the "estate") of their family corporation across generations. Fraternal polyandry is perceived and consciously selected as a means of concentrating labor and of precluding the division of a family's land and animals among its male coparceners. By virtue of this it is seen as a mechanism for maintaining or improving the wealth, power, and social status of the family. The motivation underlying the selection of fraternal polyandry is economic in nature but is concerned with wealth and social status, not subsistence survival. Tibetans do not consider fraternal polyandry a highly valued end in and of itself. . . . They can articulate quite clearly the negative aspects inherent in it as well as what, for them, are its overriding advantages. Fraternal polyandry, therefore, is not seen to be without its problems. Because authority is customarily exercised by the eldest brother, younger male siblings have to subordinate themselves with little hope of changing their status. When these younger brothers are aggressive and individualistic, intersibling tensions and difficulties often occur. Similarly, tension in polyandrous families may derive from the relationship between the wife and her husband or from the brothers' relationship concerning access to the wife. . . . Thus, while polyandry provides an answer to one type of culturally perceived problem (albeit one which the subjects see as critical), it does generate other types of problems, and the choice facing all younger male siblings is whether to trade personal freedom (monogamy) for real or potential economic security, affluence, and social prestige (fraternal polyandry). Siblings with some reservations about marrying polyandrously must assess their potential for attaining satisfactory income and social status within some reasonable period.[19]

From the perspective of participating males, there is a theoretical disadvantage to polyandry, which no doubt contributes overwhelmingly to its rare occurrence in human populations: The probability that the genes of any man will be transmitted to the next generation are lower in polyandrous marriages than in monogamous unions. Moreover, to the extent that polyandrous marriages leave some women reproductively inactive, a similar argument can be made for females. Demographic data also show lower survivorship of offspring in Tibetan fraternal polyandrous marriages than in monogamous pairings, but the explanation is not clear.

These are the facts. What then of the conclusions? As Tibetan fraternal polyandry seems to reduce the fitness of those who participate, the anthropologists Beall and Goldstein suggest "that sociocultural, economic, and political factors can perpetuate mating systems that entail significant reproductive sacrifice, i.e., can perpetuate mating systems that decrease the individual and inclusive fitness of the individuals who practice them."[20] This conclusion, it seems to me, takes too narrow a view of evolutionary considerations. We have seen above that for a long-lived species that operates close to the carrying capacity of the environment, securing resources can become an important immediate goal, essential to ultimate reproductive success. It is therefore difficult to understand why many anthropologists assume that decisions about "sociocultural, economic, and political factors" have no relevance to fitness. From the description of the conditions under which Tibetan polyandry is adopted and the reasons given by those who practice it, it clearly represents a cultural expedient lying within the behavioral repertory of the species and employed in pressing ecological circumstances. The perception of those Tibetans who opt for fraternal polyandry is that it is in the long-term interests of the family, and based only on anthropologists' head counts of the next generation, it is difficult to assert that the Tibetans are wrong. Over most of the evolutionary history of *Homo sapiens* it seems a safe bet that access to and control of material resources has been important in maximizing inclusive fitness, perhaps initially simply as a buffer against natural disasters or in competition between neighboring groups, but subsequently as economic factors of various sorts came to influence access to mates. Furthermore, if control of resources has been evolutionarily important, it is not only reasonable but in all likelihood inevitable that human nature should have come to include the necessary psychological enabling mechanisms. Why else should greed be so prevalent a force in human behavior and so potentially disruptive of society that it has to be controlled by rules? If access to resources fills a psychic need for many people, there is a good evolutionary reason. There is, in short, every reason why we should expect to find interactions between reproductive practices on the one hand and the economy and ecology of the culture on the other. Moreover, as society becomes more complex it is increasingly difficult for individuals to act in ways that are unambiguously in the best interests of their genes. They will be confronted with competing motivations—the proximate enabling mechanisms for different elements in the panoply of behaviors that collectively ensure reproductive success and survival. Many times, as in the case of Tibetans, they will not have all the information they need. Far from providing a negative test of sociobiological theory, the low incidence of polyandry, the conditions under which it appears, and the conflicting impulses of the Tibetans who practice it, all seem quite compatible with what one would expect of a highly intelligent, resource-limited mammal.

Evolution can help us to understand ourselves, but in a world of complex technology many of our actions may seem at first glance to be decoupled from that history. If access to resources and attention to one's well-being were critical to reproductive success during most of our evolutionary history, as much of our behavior suggests is the case, the proximate enabling mechanisms that impel behavior to those ends can be so powerful that they begin to take on a life of their own. When the genie gets out of the bottle, he may call his own tune. The demographic transition is a phenomenon in which individuals in post-industrial societies are quite content to have fewer children in order to invest more in themselves and to enjoy the good life. Is this inconsistent with everything this book is about? I think not. Many people in affluent societies also eat more than is good for themselves, drink alcohol to the extent that they jeopardize their careers if not their lives, imperil themselves with mind-altering drugs, and engage in dangerous behavior such as jumping out of airplanes, climbing cliffs, or simply driving automobiles too fast, all to satisfy psychological urges that may not only be divorced from the goal of maximizing fitness, but may actually jeopardize it. A number of powerful psychic forces propelling self-gratification may be miscast in the contemporary world, but they made their evolutionary debut in another theater and they continue to play the only roles they know.

9

Epilogue—Concerning "Biological Reductionism"

It has become fashionable to dismiss evolutionary thinking as "biological reductionism" (see, for example, the quotation from Marvin Harris in the previous chapter). The underlying premise is that cultural anthropology and sociology deal with complex levels of organization that exhibit emergent properties, and the concepts appropriate for description and analysis at lower levels are inadequate when applied to the social sciences. Attributing behavior to genes is therefore seen as an inappropriate effort at reductionism.

This view contains a kernel of truth but is frequently incorrect in its application. We have considered at some length the mischief that flows from a naive understanding of the interplay of genetic and epigenetic events in determining behavior. I will not belabor that point further. I suggest, however, that there is still another area of misunderstanding that needs the spotlight directed on it. In general, the interfaces between academic disciplines in science are defined by emergent properties, and in this respect the line that has been drawn between the social and natural sciences is not unique. The field of chemistry exists to understand the stuff of which the world of our senses is made. This stuff, of course, consists of electrons and the elementary particles of atomic nuclei, but one can know a lot about elementary particles and not be adequately prepared to understand why CO_2 is a gas and SiO_2 is a rock. Similarly, the self-replicating chemical systems of cells represent another level of molecular organization with their own distinct properties.

Nevertheless, it is equally true that an understanding of chemical bonding in terms of atomic structure brings order to chemistry, and the enormous advances in this century in biochemistry and molecular biology would not have been possible without an appreciation of the various kinds of chemical bonds that exist in cells. The study of the chemistry of life was destined to flounder about in a hopeless fashion when our anchoring concepts were limited to chemically undefined colloids, coacervates, and soups. To be able to write on paper the primary structure (the sequence of amino acids) of a protein molecule is an organizing principle of the greatest importance, but it will not, in itself, show us the shape of the molecule in aqueous solution or how the molecule interacts with a substrate in enzymatically catalyzing a critical metabolic reaction. The latter are properties that follow from a more complex level of organization, but that, in their turn, can only be understood in a context that is consistent with, and incorporates the concepts of, the rest of physics and chemistry.

A similar relationship exists between the social sciences and biology. During the last thirty years, knowledge of the nervous system has increased at an ever-accelerating pace, and it is possible to identify a couple of its organizing principles. First, the dynamic properties of the nervous system are based on the propagation of impulses along nerve fibers and the interactions between cells that occur at structures called synapses. In a sense, synapses are analogous to chemical bonds, and not so many years ago many of us thought we understood the rules by which synapses work. Today it is clear that we understood only some of the rules; the discovery of neuro-active peptides and slowly acting synaptic transmitters and modulators makes it clear we still have a distance to travel. A second organizing principle of the nervous system is that in general particular functions are handled by specific populations of cells. This fact was obscured for a number of years because of the existence of parallel processing and other anatomical complexities, but today most neurobiologists enter the laboratory in the morning believing that the complex functions of higher neural circuits that generate behavior, including the mental activities of the human brain, will ultimately be understood in terms of (at the very least!) these two general principles.

All of this has to do with proximate cause. There have been some very interesting and parallel theoretical advances in evolutionary biology during the past twenty years that address matters of ultimate cause and that are of equally potential importance for the social sciences. By this I do not mean that anthropology, human psychology, and sociology will "reduce to biology" any more than biology has reduced to chemistry. Each level of organization is presented with properties that derive from the fact of organization and that have to be studied on their own terms. But those terms must relate to the rest of science or the exercise will be of no lasting value. The evolutionary concepts deal not primarily with physiology and chemistry but with biology's historical dimension: evolution. Evolution enters science with biology, but it is an exercise in self-deception to suppose that it leaves the cast when the social sciences come onto the stage. Cognitive psychologists certainly understand that their explanatory models must not only have heuristic value at their own descriptive levels, but, in some final sense, must also relate to the functions of neurons. Similarly, a larger problem facing the social sciences is to generate a theoretical architecture that neither clashes with biology nor consists of such exclusive structures that it is ultimately rendered irrelevant. "Biological reductionism" can be used to trivialize evolutionary biology in an effort to keep it at arm's length, but this is an unworthy goal. The social sciences will have matured only when they are firmly grounded in, and consistent with, the rest of our understanding of nature. In this enterprise, biology and the social sciences should work together, for the discovery of principles that can unite hierarchies and cut across species will enrich our knowledge and our lives. To date, only evolutionary biology offers that framework.[1]

Notes and References

The following notes and references will help to open the literature on this diverse subject. I have made no special effort to refer to original articles in scientific journals. Most of the citations are to books or reviews that themselves have extensive bibliographies. Many of the references can therefore be pursued without access to a major research library.

Chapter 1

1. Foreword to Dawkins, R. *The Selfish Gene*. Oxford University Press, NY (1976).
2. Nelkin, D., "The science-textbook controversies." *Scientific American* 234:33-39 (1976).
3. Schafersman, S., "Censorship of Evolution in Texas." *Creation/ Evolution* 3:30-34 (1982).
4. Dawkins, *op. cit.*
5. Mayr, E. Prologue: Some thoughts on the history of the evolutionary synthesis. In E. Mayr and W.B. Provine, eds., *The Evolutionary Synthesis: Perspectives on the Unification of Biology,* pp 1-48. Harvard University Press, Cambridge, MA (1980).
6. For the history of this terminology see Hailman, J.P., Ontogeny: Toward a general theoretical framework for ethology. In P.P.G. Bateson and P.H. Klopfer, eds., *Perspectives in Ethology, 5: Ontogeny,* pp 133-189. Plenum, NY (1982).
7. Mayr. *op.cit.*
8. Westermarck, E. *The Origin and Development of the Moral Ideas,* Vol II. Macmillan, London (1917).
9. Freud, S. *Totem and Taboo.* In A.A. Brill, ed. and trans., *The Basic Writings of Sigmund Freud.* Random House (Modern Library edition), NY (1938).
10. Shepher, J., "Mate selection among second generation kibbutz adolescents and adults: Incest avoidance and negative imprinting." *Archives of Sexual Behavior* 1:293-307 (1971).

143

11. Wolf, A.P., "Childhood association, sexual attraction and the incest taboo: A Chinese case." *American Anthropologist* 68:883-898 (1966).
12. Bateson, P.P.G. Rules for changing the rules. In Bendall, D.S., ed. *Evolution from Molecules to Men.* Cambridge University Press, Cambridge, England (1983) pp 483-507. This paper should be consulted not only for its discussion of work on mate choice in animals but for a critical review of the data on humans referred to in the two preceding notes.

Chapter 2

1. Lerner, L.S. and W.J. Bennetta. "The treatment of theory in textbooks." *Science Teacher,* April:37-41 (1988).
2. Committee on Science and Creationism: *Science and Creationism: A View from the National Academy of Sciences.* National Academy Press, Washington, DC (1984) 28 pp. This brief account, produced under the imprimatur of the National Academy of Sciences, was prompted by the litigation over the teaching of creationism as science. It is a concise and readable statement about the scientific basis of evolution.
3. Bishop, J.A. and L.M. Cook., "Industrial melanism and the urban environment." *Advances in Ecological Research* 11:373-404 (1980).
4. Dayhoff, M.O., ed. *Atlas of Protein Sequence and Structure.* National Biomedical Research Foundation, Washington, DC (1972 and subsequent supplements). For a more general account of proteins see Creighton, T.E. *Proteins: Structures and Molecular Principles.* W.H. Freeman and Co., NY (1984).
5. Bendall, D.S., ed. *Evolution from Molecules to Men.* Cambridge University Press, Cambridge, England (1983). This symposium, commemorating the centenary of Charles Darwin's death, brought together an unusually broad spectrum of distinguished evolutionary biologists.
6. Committee on Science and Creationism, *op. cit.*
7. Gould, S.J. *Wonderful Life: The Burgess Shale and the Nature of History.* W.W. Norton & Co., NY (1989). Among other things, a detailed account of numerous invertebrate groups that are now extinct and that were utterly unknown in Darwin's day.
8. Gingerich, P.D., B.H. Smith, and E.L. Simons, "Hind limbs of Eocene *Basilosaurus*: Evidence of feet in whales." *Science* 249:154-157.
9. Denton, M. *Evolution: A Theory in Crisis.* Adler and Adler, Bethesda, MD (1985). Although it carefully eschews creationist arguments, this book is used as ammunition for their popguns.
10. Dawkins, R. *The Blind Watchmaker: Why the evidence of evolution reveals a universe without design.* W.W. Norton & Co., NY (1986). An engaging account for the lay reader.
11. *Ibid.*

12. Denton, *op. cit.*

13. Bateson, W., "Review of the Mechanism of Mendelian Heredity, by T.H. Morgan, A.H. Sturtevant, H.J. Muller and C.B. Bridges." *Science* 44:536–543 (1916).

Chapter 3

1. Selander, R.K. Genic variation in natural populations. In F.J. Ayala, ed., *Molecular Evolution,* Sinauer Associates, Sunderland, MA (1976), pp 21–45.

2. Alberts, B., D. Bray, J. Lewis, M. Raff, K. Roberts, and J.D. Watson. *Molecular Biology of the Cell.* 2nd ed. Garland Publishing Co., NY (1989). A massive but clear account; very readable in short stretches and a gem to consult.

3. Doolittle, W.F. and C. Sapienza, "Selfish genes, the phenotype paradigm and genome evolution." *Nature* 284:601–603 (1980). Orgel, L.E. and F.H.C. Crick, "Selfish DNA: the ultimate parasite." *Nature* 284:604–607 (1980).

4. For a general textbook on evolutionary principles see, for example, Futuyma, D.J. *Evolutionary Biology,* 2nd ed. Sinauer Associates, Inc., Sunderland, MA (1986).

5. Gould, S.J. and E.S. Vrba, "Exaptation—a missing term in the science of form." *Paleobiology* 8:4–15 (1982).

6. *Ibid.*

7. Endler, J.A., "Natural selection of color patterns in *Poecilia reticulata.*" *Evolution* 34:76–91 (1980).

8. Land, M.F. Optics and vision in invertebrates. In H. Autrum, ed., *Handbook of Sensory Physiology, VII/6B: Invertebrate Visual Centers.* Springer Verlag, Berlin (1981) pp 472–592.

9. Gould, S.J. and R.C. Lewontin. The spandrels of San Marco and the Panglossian paradigm: a critique of the adaptationist programme. *Proceedings of the Royal Society of London,* Series B, 205:581–598 (1979).

10. Goldsmith, T.H., "Optimization, constraint, and history in the evolution of eyes." *Quarterly Review of Biology* 65:281–322 (1990).

11. *Ibid.*

12. The reconciliation of apparently altruistic behavior with the genetic selfishness required by natural selection effectively began with the insights of W.D. Hamilton. See, for example his papers: "The genetical theory of social behaviour," I,II, *Journal of Theoretical Biology* 7:1–52 (1964); "Selfish and spiteful behaviour in an evolutionary model," *Nature,* London 228:1218–1220 (1970); "Geometry for the selfish herd," *Journal of Theoretical Biology* 31:295–311 (1971); "Selection of selfish and altruistic behavior in some extreme models." In J.F. Eisenberg

and W.S. Dillon, eds., *Man and Beast: Comparative Social Behavior*, pp 57–91, Smithsonian Institution Press, Washington, DC (1971); "Altruism and related phenomena, mainly in social insects," *Annual Review of Ecology and Systematics* 3:193–232 (1972).

13. Sherman, P.W. The limits of ground squirrel nepotism, in G.W. Barlow and J. Silverberg, eds., *Sociobiology: beyond nature/nurture?* Westview Press, Boulder, CO (1980).

14. Dawkins, The Selfish Gene, *op.cit.*

15. Williams, G.C. *Adaptation and Natural Selection: A Critique of Some Current Evolutionary Thought.* Princeton University Press, Princeton, NJ (1966).

16. For a general discussion with references to mathematical models see pp 106–117 in E.O. Wilson, *Sociobiology: The New Synthesis.* The Belknap Press of Harvard University Press, Cambridge, MA (1975).

17. Doolittle and Sapienza, *op.cit.*

18. Buss, L. *The Evolution of Individuality.* Princeton University Press, Princeton, NJ (1987).

19. For contrasting views see: Ayala, F.J., Microevolution and macroevolution, pp 387–402, and Gould, S.J., Irrelevance, submission and partnership: the changing role of palaeontology in Darwin's three centennials and a modest proposal for macroevolution, pp 347–366, both in Bendall, D.S., ed., *Evolution from Molecules to Men.* Cambridge University Press, Cambridge, England (1983). (See also note 5 to Chapter 2.)

20. LeBoeuf, B.J. and R.S. Peterson, "Social status and mating activity in elephant seals." *Science* 163:91–93 (1969).

21. Trivers, R.L., "The evolution of reciprocal altruism." *Quarterly Review of Biology* 46: 35–57 (1971). Trivers, R.L. "Parent-offspring conflict." *American Zoologist* 14:249–264 (1974). Trivers, R. *Social Evolution.* Benjamin/Cummings Publishing Company, Inc., Menlo Park, CA (1985).

22. Maynard Smith, J., "Optimization theory in evolution." *Annual Review of Ecology and Systematics.* 9:31–56 (1978). Maynard Smith, J., "The evolution of behavior." In *Evolution,* pp 92–101. Reprints from *Scientific American,* September 1978.

23. For a more detailed discussion of the regulation of clutch size in birds see pp 25–28 in Daly, M. and M. Wilson, *Sex, Evolution, and Behavior.* Wadsworth Publishing Co., Belmont, CA (1983) 2nd ed.

Chapter 4

1. See notes 15 and 16 for Chapter 3.

2. See note 12 for Chapter 3.

3. See note 21 for Chapter 3.

4. See note 22 for Chapter 3.

5. Van Valen, L., "A new evolutionary law." *Evolutionary Theory* 1:1–30 (1973).

6. Trivers, R. *Social Evolution.* Benjamin/Cummings Publishing Company, Inc., Menlo Park, CA (1985).

7. Michod, R.E. and B.R. Levin, eds. *The Evolution of Sex.* Sinauer Associates, Sunderland, MA (1988).

8. Dawkins, The Selfish Gene. *op. cit.*

9. Daly, M. and M. Wilson. *Sex, Evolution, and Behavior.* Wadsworth Publishing Co., Belmont, CA (1983) 2nd ed. This book provides a good account of the reproductive diversity of animals and its evolutionary significance.

10. Dawkins, *The Selfish Gene, op. cit.*

11. For a general description and references to work on macaques see pg 294 in Wilson, E.O. Sociobiology: The New Synthesis. The Belknap Press of Harvard University Press, Cambridge, MA (1975). Early observations on chimpanzees were described in a popular account by Goodall, J.v.L. *In the Shadow of Man,* Dell, NY (1971).

12. Hrdy, S.B. *The Woman That Never Evolved.* Harvard University Press, Cambridge, MA (1981).

13. Barash, D. *The Whisperings Within: Evolution and the Origin of Human Nature.* Penguin Books, NY (1979).

14. Dawkins, *The Selfish Gene, op. cit.*

15. Daly and Wilson, *op. cit.*

16. *Ibid.*

17. Barash, *op. cit.*

18. R.W. Wrangham. Doctoral Thesis, Cambridge University (1975), cited in Daly and Wilson (note 9).

19. Daly and Wilson, *op. cit.*

20. Stearns, S.C., "Life-history tactics: a review of the ideas." *Quarterly Review of Biology* 51:3–47 (1976). Stearns, S.C., "The evolution of life history traits: a critique of the theory and a review of the data." *Annual Review of Ecology and Systematics* 8:145–171 (1977).

21. Harris, M. *Cultural Materialism.* Random House, NY (1979).

22. Murdock, G.P. *Ethnographic Atlas.* University of Pittsburgh Press, Pittsburgh, PA (1967). See also notes 9 and 12.

23. Lee, R.B. The !Kung bushmen of Botswana. In M.G. Bicchieri, ed., *Hunters and Gatherers Today.* Holt, Rinehart and Winston, NY (1972).

24. Betzig, L.L. *Despotism and Differential Reproduction: A Darwinian View of History.* Aldine De Gruyter, Hawthorne, NY (1986).

25. See pages 280-281 in Daly and Wilson (note 9) for references and further discussion.

26. The somewhat different ways the Coolidge effect has been defined and the variations in degree of expression are discussed by Dewsbury, D.A., "Effects of novelty on copulatory behavior: the Coolidge effect and related phenomena," *Psychological Bulletin* 89:464-482 (1981). For general information consult Symons, D., *The Evolution of Human Sexuality,* Oxford University Press, NY (1979).

27. See pages 295-297 in Daly and Wilson (note 9).

28. Daly, M. and M. Wilson. *Homicide.* Aldine De Gruyter, Hawthorne, NY (1988).

29. Daly and Wilson, Homicide, *op. cit.*

30. Thornhill, R. and N.W. Thornhill, "Human rape: An evolutionary analysis." *Ethology and Sociobiology* 4:137-173 (1983).

31. Daly and Wilson, *Homicide, op. cit.*

Chapter 5

1. As translated by E.S. Russell, pg 35, *The Interpretation of Development and Heredity. A Study in Biological Method,* Clarendon Press, Oxford (1930). As Oppenheim (see note 7) points out, the "essential nature" to which von Baer referred was more a philosophical than a scientific concept and drew on the influence of Goethe. One can give von Baer's words a contemporary interpretation, however, if the "essential nature" is taken to mean the overall design of the organism that has been forged by natural selection and that has a representation in the information carried by the genes.

2. Freeman, D. Sociobiology: The "antidiscipline" of anthropology. In A. Montagu, ed., *Sociobiology Examined,* pp 198-219. Oxford University Press, NY (1980).

3. Dawkins, The Selfish Gene. *op. cit.*

4. Freeman, *op. cit.*

5. Beach, F.A., "The descent of instinct." *Psychological Review* 62:401-409 (1955).

6. For example, Konishi, M., "The role of auditory feedback in the control of vocalization in the white crowned sparrow." *Zeitschrift für Tierpsychologie* 22:770-783 (1965).

7. Oppenheim, R.W. Preformation and epigenesis in the origins of the nervous system and behavior: issues, concepts, and their history. In P.P.G. Bateson and P.H. Klopfer, eds., *Perspectives in Ethology,* pp 1-100. Plenum Press, NY (1982).

8. Stern, C.D. and R.J. Keynes, "Spatial patterns of homeobox gene expression in the developing mammalian CNS." *Trends in Neuroscience* 11:190-191 (1988).

9. For an overview of the development of the nervous system consult Chapters 55 and 56 in Kandel, E. and J.H. Schwartz, *Principles of Neural Science,* 2nd ed., Elsevier, NY (1985).

10. For a general account see Kelly, D.D., Sexual differentiation of the nervous system, Chapter 58 in Kandel, E. and J.H. Schwartz, *Principles of Neural Science,* 2nd ed., Elsevier, NY (1985), or Sexual development and differentiation, Chapter 10 in Daly, M. and M. Wilson, *Sex, Evolution, and Behavior,* Wadsworth Publishing Co., Belmont, CA (1983) 2nd ed.

11. Goldman, Ari L., "Religion Notes," *The New York Times,* July 28, 1990.

12. Naftolin, F. and H.L. Judd, "Testicular feminization," in R. Wynn, ed., *Ob-Gyn Annual* (1973), pp 25-53.

13. Gorski, R.A., J.H. Gordon, J.E. Shryne, and A.M. Southam, "Evidence for a morphological sex difference within the medial preoptic area of the rat brain." *Brain Research* 148:333-346 (1978).

14. Raisman, G. and P.M. Field, "Sexual dimorphism in the preoptic area of the rat." *Science* 173:731-733 (1971).

15. Goldman, P.S., H.T. Crawford, L.P. Stokes, T.W. Galkin, and H.E. Rosvold, "Sex-dependent behavioral effects of cerebral cortical lesions in the developing rhesus monkey." *Science* 186:540-542 (1974).

16. Witelson, S.F., "Sex and the single hemisphere: specialization of the right hemisphere for spatial processing." *Science* 193:425-427 (1976).

17. An extensive account of this Nobel Prize-winning work is available in a nicely illustrated book: Hubel, D. *Eye, Brain, and Vision.* Scientific American Books, distributed by W.H. Freeman and Co., NY (1988).

18. Harlow, H.F. and M.K. Harlow, "Social deprivation in monkeys." *Scientific American* 207:137-146 (1962).

19. Spitz, R.A., "Hospitalism: an inquiry into the genesis of psychiatric conditions in early childhood." *Psychoanalytic Study of the Child* 1:53-74 (1945). Spitz, R.A., "Hospitalism." A follow-up report on investigation described in Vol 1, 1945. *Psychoanalytic Study of the Child* 2:113-117 (1946). Spitz, R.A. and K.M. Wolf, "Anaclitic depression: An inquiry into the genesis of psychiatric conditions in early childhood," II. *Psychoanalytic Study of the Child* 2:313-342 (1946). Wolf, A.P., "Childhood association, sexual attraction and the incest taboo: A Chinese case." *American Anthropologist* 68:883-898 (1966).

20. Lenneberg, E.H. *Biological Foundations of Language.* John Wiley and Sons, NY (1967).

21. Thompson, R.F., T.W. Berger, and J. Madden, IV, "Cellular processes of learning and memory in the mammalian CNS." *Annual Review of Neuroscience* 6:447-491 (1983).

22. Kandel, E.R. and J.H. Schwartz., "Molecular biology of learning: modulation of transmitter release," *Science* 218:433-443 (1982). Hawkins, R.D. and E.R. Kandel, "Is there a cell-biological alphabet for simple forms of learning?" *Psychological Review* 91:375-391 (1984).

23. Kandel, E.R. Environmental determinants of brain architecture and of behavior: Early experience and learning, in E.R. Kandel and J.H. Schwartz, *Principles of Neural Science*, 1st ed., Elsevier, NY (1981).

24. Bateson, G. *Mind and Nature, a Necessary Unity*. Dutton, NY (1979).

25. Freedman, D.G., "Smiling in blind infants and the issue of innate vs. acquired." *Journal of Child Psycholology and Psychiatry* 5:171-184 (1964). Eibel-Eibesfeldt, I. *Ethology: the Biology of Behavior*. Holt, Rinehart and Winston, NY (1970). Eibel-Eibesfeldt, I. *Human Ethology*. Aldine de Gruyter, NY (1989).

26. Lumsden, C.J. and E.O. Wilson. *Genes, Mind and Culture*. Harvard University Press, Cambridge, MA (1981).

27. Bonner, J.T. *The Evolution of Culture in Animals*. Princeton University Press, Princeton, NJ (1980).

28. Williams, G.C. "Pleiotropy, natural selection, and the evolution of senescence." *Evolution* 11:398-411 (1957).

Chapter 6

1. Darwin, C. *The Descent of Man and Selection in Relation to Sex*. 2nd ed., pg 126, D. Appleton and Company, NY (1880).

2. Alexander, R.D., "The search for an evolutionary philosophy of man." *Proceedings of the Royal Society of Victoria* 84:99-120 (1971).

3. For a brief review and references, see pp 146-147 in Wilson, *op. cit.* in Chapter 3, note 16.

4. *Ibid.*, pp 147-149 and 514-515, note 3.

5. Seligman, M.E.P., "On the generality of the laws of learning." *Psychological Review* 77:406-418 (1970).

6. Azrin, N.H., "Some notes on punishment and avoidance." *Journal of the Experimental Analysis of Behavior* 2:260 (1959).

7. Konorski, J. *Integrative Activity of the Brain*. University of Chicago Press, Chicago (1967).

8. Thorndike, E.L. *Animal Intelligence*. Hafner, NY (1965). (Originally published: Macmillan, NY (1911).)

9. Delius, J.D. and J. Emmerton. Visual performance of pigeons. In A.M. Granda and J.H. Maxwell, eds., *Neural Mechanisms of Behavior in the Pigeon*, pp 51-70. Plenum, NY (1979).

10. Wilson, *op. cit.*

11. Shettleworth, S.J. Constraints on learning. In S. Lehrman, R.A. Hinde, and E. Shaw, eds., *Advances in the Study of Behavior,* pp 1-68. Academic Press, NY (1972). Hinde, R.A. and J. Stevenson-Hinde, eds. *Constraints on Learning: Limitations and Predispositions,* Academic Press, NY (1973).

12. Garcia, J., B.K. McGowan, F.R. Ervin, and R.A. Koelling, "Cues: their relative effectiveness as a function of the reinforcer." *Science* 160:794-795 (1968).

13. Wilcoxon, H.C., W.B. Dragoin, and P.A. Kral. "Illness-induced aversions in rat and quail: relative salience of visual and gustatory cues." *Science* 171:826-828 (1971).

14. Rozin, P. and J.W. Kalat. "Specific hungers and poison avoidance as adaptive specializations of learning." *Psychological Review* 78:459-486 (1971).

15. Krane, R.V. and A.R. Wagner. "Taste aversion learning with a delayed shock US: Implications for the 'Generality of the Laws of Learning.'" *Journal of Comparative and Physiological Psychology* 88:882-889 (1975).

16. Gillette, K., G.M. Martin, and W.P. Bellingham. "Differential use of food and water cues in the formation of conditioned aversions by domestic chicks (*Gallus gallus*)." *Journal of Experimental Psychology: Animal Behavior Processes* 6:99-111 (1980).

17. Goldsmith, T.H., J.S. Collins, and D.L. Perlman. "A wavelength discrimination function for the hummingbird *Archilochus alexandri.*" *Journal of Comparative Physiology A,* 143:103-110 (1981).

18. Mackintosh, N.J. Stimulus selection: learning to ignore stimuli that predict no change in reinforcement. In R.A. Hinde and J. Stevenson, eds., *Constraints on Learning: Limitations and Predispositions, op.cit.* in this chapter, note 11.

19. Wilson, *op. cit.* Sebeok, T.A., ed. *How Animals Communicate.* Indiana University Press, Bloomington (1977).

20. Griffin, D.R. *The Question of Animal Awareness: Evolutionary Continuity of Mental Experience.* Rockefeller University Press, NY (1981).

21. Gardner, B.T. and R.A. Gardner, "Teaching sign language to a chimpanzee," *Science* 165:664-672 (1969); Two-way communication with an infant chimpanzee, in A.M. Schrier and F. Stollnitz, eds., *Behavior of Non-Human Primates,* Vol IV, Chapter 3, Academic Press, NY (1971); "Evidence for sentence constituents in the early utterances of child and chimpanzee," *Journal of Experimental Psychology* 104:244-267 (1975). Fouts, R.S., W. Chown, and L. Goodwin, "Translation from vocal English to American Sign Language in a chimpanzee (Pan)," *Learning and Motivation* 7:458-475

(1976). Premack, D., *Intelligence in Ape and Man,* Erlbaum, Hillsdale, NJ (1976). Rumbaugh, D.M., *Language Learning by a Chimpanzee,* Academic Press, NY (1977). Savage-Rumbaugh, E.S., D.M. Rumbaugh, and B. Boysen, "Linguistically mediated tool use and exchange by chimpanzees (*Pan troglodytes*)," *Behavioral and Brain Sciences* 4:539-554 (1978); "Do apes use language?," *American Scientist* 68:49-61 (1980). Terrace, H., L.A. Petitto, R.J. Sanders, and T.G. Bever, "Can an ape create a sentence?" *Science* 206:891-902 (1979).

22. Pepperberg, I.M., "Cognition in the African Grey parrot: preliminary evidence for auditory/vocal comprehension of the class concept," *Animal Learning and Behavior* 11:179-185 (1983); "Evidence for conceptual quantitative abilities in the African grey parrot: labeling of cardinal sets," *Ethology* 75:37-61 (1987).

23. Griffin, *op. cit.*

24. Seyfarth, R.M., D.L. Cheyney, and P. Marler, "Monkey responses to three different alarm calls: evidence of predator classification and semantic communication," *Science* 210:801-803 (1980); "Vervet monkey alarm calls: semantic communication in a free-ranging primate," *Animal Behavior* 28:1070-1094 (1980).

25. Griffin, *op. cit.*

26. Frisch, K. von. *The Dance Language and Orientation of Bees,* trans. by L. Chadwick, Harvard University Press, Cambridge, MA (1967). See also D.R. Griffin, note 18.

27. Lindauer, M. *Communication Among Social Bees,* Harvard University Press, Cambridge, MA (1961). Seeley, T. *Honeybee Ecology: A Study of Adaptation in Social Life,* Princeton University Press, Princeton, NJ (1985).

28. Chomsky, N. *Cartesian Linguistics.* Harper and Row, NY (1966).

Chapter 7

1. Tuchman, B.W. *The March of Folly: From Troy to Vietnam.* Knopf, distributed by Random House, NY (1984).

2. MacLean, P.D. Clarence M. Hinks Memorial Lectures, 1969. In T.G. Boag and D. Campbell, eds., *A Triune Concept of the Brain and Behavior.* University of Toronto Press, Toronto (1973). For a general functional description of the human brain see Kandel, E. and J.H. Schwartz, *Principles of Neural Science,* 2nd ed., Elsevier, NY (1985).

3. Russel, I.S. Brain size and intelligence: a comparative perspective. In D.A. Oakley and H.C. Plotkin, eds., *Brain, Behavior and Evolution,* pp 126-153. Methuen, London (1979).

4. Oakley, D.A. Cerebral cortex and adaptive behavior. In D.A. Oakley and H.C. Plotkin, *ibid.*

5. Griffin, D.R. *The Question of Animal Awareness: Evolutionary Continuity of Mental Experience. op. cit.*

6. Stacey, P.B. and C.E. Bock, "Social plasticity in the acorn woodpecker." *Science* 202:1298–1300 (1978).

7. Wilson, *op. cit.,* p 20.

8. Reviewed by Griffin, D.R. *Animal Thinking.* Harvard University Press, Cambridge, MA (1984).

9. Kamil, A.C. and T.D. Sargent. *Foraging Behavior: Ecological, Ethological and Psychological Approaches.* Garland STPM Press, NY (1981).

10. Lewontin, R.C. Fitness, survival and optimality. In D.J. Horn, G.R. Stairs and R.D. Mitchell, eds., *Analysis of Ecological Systems,* pp 3–21. Ohio State University Press, Columbus (1977).

11. Maynard Smith, *op. cit.,* Chapter 3, note 22.

12. *Ibid.* and see also Alexander, R.D., "The search for a general theory of behavior," *Behavioral Science* 20:77–100 (1975).

13. Wilson, E.O. *On Human Nature.* Harvard University Press, Cambridge, MA (1978).

14. Gould, S.J. "A Review of 'On Human Nature' by E.O. Wilson," *Human Nature* 1(10):20–28 (1978).

15. Daly and Wilson, *Homicide, op. cit.*

16. Alexander, R.D. *The Biology of Moral Systems.* Aldine De Gruyter, Hawthorne, NY (1987).

17. Daly and Wilson, *Homicide, op. cit.*

18. Alexander, *op. cit.,* pp 34–35; p 93

19. *Ibid.,* p 40

Chapter 8

1. For a brief account and references to the original Japanese work see pp 170–171 in E.O. Wilson, *Sociobiology: The New Synthesis, op. cit.*

2. Bonner, *op. cit.* See also Wilson, note 1.

3. Mundinger, P.C., "Animal cultures and a general theory of cultural evolution." *Ethology and Sociobiology* 1:183–223 (1980).

4. See notes 2 and 3, as well as Alexander, R.D., Evolution and culture, in N.A. Chagnon and W. Irons, eds., *Evolutionary Biology and Human Social Behavior: An Anthropological Perspective,* pp 59–78, Duxbury Press, North Scituate, MA (1979); Boyd, R. and P.J. Richerson, "A simple dual inheritance model of the conflict between social and biological evolution," *Zygon* 11:254–262 (1976); *Culture and the Evolutionary Process,* University of Chicago Press, Chicago (1985); Cloninger, C.R., J. Rice, and T. Reich, "Multifactorial inheritance with

cultural transmission and assortative mating: II, a general model of combined polygenic and cultural inheritance," *American Journal of Human Genetics*, 31:176-198 (1979); "Multifactorial inheritance with cultural transmission and assortative mating: III, family structure and the analysis of separation experiments," *American Journal of Human Genetics* 31:366-388 (1979); Durham, W.H., "The adaptive significance of cultural behavior," *Human Ecology* 4:89-121 (1976); Durham, W.H., Toward a coevolutionary theory of human biology and culture, in N.A. Chagnon and W. Irons, eds., *Evolutionary Biology and Human Social Behavior: An Anthropological Perspective,* pp 39-59, Duxbury Press, North Scituate, MA (1979); Feldman, M.W. and L.L. Cavalli-Sforza, "Cultural and biological evolutionary processes, selection for a trait under complex transmission," *Theoretical Population Biology* 9:238-259 (1976); "Aspects of variance and covariance analysis with cultural inheritance," *Theoretical Population Biology* 15:276-307 (1979); Pulliam, H.R. and C. Dunford, *Programmed to Learn: An Essay on the Evolution of Culture,* Columbia University Press, NY (1979); Rice, J., C.R. Cloninger, and T. Reich, "Multifactorial inheritance with cultural transmission and assortative mating: I, description and basic properties of the unitary models," *American Journal of Human Genetics* 30:618-643 (1978).

5. Mundinger, *op. cit.*

6. *Ibid.*

7. *Ibid.*

8. The structuralist mode of analysis characterizes cultures in terms of what are assumed to be basic (structural) components, such as systems of kinship and mythology. Levi-Strauss, C., Structural Anthropology, Basic Books, NY (1963); Laughlin, C.D. and E.G. d'Aquili, Biogenetic Structuralism, Columbia University Press, NY (1974).

9. Lumsden and Wilson, *Genes, Mind and Culture. op. cit.,* Chapter 5, note 26.

10. Harris, *op. cit.,* Chapter 4, note 21.

11. *Ibid.,* pg 56.

12. *Ibid.,* pg 63.

13. *Ibid.,* pg 121.

14. *Ibid.,* pg 126.

15. *Ibid.,* pp 139-140.

16. Meddis, R. The evolution and function of sleep. In D.A. Oakley and H.C. Plotkin, eds., *Brain, Behavior and Evolution,* pp 99-125. Methuen, London (1979)

17. *Ibid.*

18. Beall, C.M. and M.C. Goldstein, "Tibetan fraternal polyandry: a test of sociobiological theory." *American Anthropologist* 83:5–12 (1981).
19. *Ibid.*
20. *Ibid.*

Chapter 9

1. This theme has been developed at greater length by Wilson, E.O., "Biology and the social sciences," *Daedalus* 106:127–140 (1977).

Index